VOLUME NINETY NINE

ADVANCES IN GENETICS

Natural Variation and Clocks

ADVANCES IN GENETICS, VOLUME 99

Serial Editors

Theodore Friedmann
Department of Pediatrics, University of California at San Diego, School of Medicine, CA, USA

Jay C. Dunlap
Department of Molecular and Systems Biology, The Geisel School of Medicine at Dartmouth, Hanover, NH, USA

Stephen F. Goodwin
Centre for Neural Circuits and Behaviour, University of Oxford, Oxford, UK

VOLUME NINETY NINE

ADVANCES IN GENETICS

Natural Variation and Clocks

Edited by

STEPHEN F. GOODWIN
Centre for Neural Circuits and Behaviour,
University of Oxford, Oxford, UK

ELSEVIER

ACADEMIC PRESS
An imprint of Elsevier

Academic Press is an imprint of Elsevier
50 Hampshire Street, 5th Floor, Cambridge, MA 02139, United States
525 B Street, Suite 1800, San Diego, CA 92101–4495, United States
The Boulevard, Langford Lane, Kidlington, Oxford OX5 1GB, United Kingdom
125 London Wall, London, EC2Y 5AS, United Kingdom

First edition 2017

ISBN: 978-0-12-811811-5
ISSN: 0065-2660

For information on all Academic Press publications
visit our website at https://www.elsevier.com/books-and-journals

Publisher: Zoe Kruze
Acquisition Editor: Zoe Kruze
Editorial Project Manager: Alina Cleju
Production Project Manager: James Selvam
Cover Designer: Mark Rogers

Typeset by SPi Global, India

CONTENTS

CONTRIBUTORS

William E. Bradshaw
Institute of Ecology and Evolution, University of Oregon, Eugene, OR, United States

Christina M. Holzapfel
Institute of Ecology and Evolution, University of Oregon, Eugene, OR, United States

Bala S.C. Koritala
Department of Biology; Center for Computational and Integrative Biology, Rutgers, The State University of New Jersey, Camden, NJ, United States

Kwangwon Lee
Department of Biology; Center for Computational and Integrative Biology, Rutgers, The State University of New Jersey, Camden, NJ, United States

Malcolm von Schantz
Faculty of Health and Medical Sciences, University of Surrey, Guildford, Surrey, United Kingdom

PREFACE

Characterizing the genetic and molecular mechanisms underlying natural variation of circadian clocks remains a challenging task in any organism. Thematic Volume 99 contains three comprehensive reviews dealing with these issues of Natural variation and clocks.

Chapter 1 assesses the important roles fungi play in ecosystems on earth and explores how fungi, such as *Neurospora,* can serve as an important model for understanding the ecological role of natural variation in fungal circadian clocks. Chapter 2 describes how a small mosquito, *Wyeomyia smithii,* that lives only in the water-filled leaves of a carnivorous plant over a wide geographic range, can be used to explore the genetic and evolutionary basis of photoperiodism. Chapter 3 looks at the evidence for Natural variation in circadian rhythms and sleep in humans.

I am most grateful to the authors for their excellent reviews.

STEPHEN F. GOODWIN
Oxford, UK

CHAPTER ONE

Natural Variation of the Circadian Clock in *Neurospora*

Bala S.C. Koritala*,†, Kwangwon Lee*,†,1
*Department of Biology, Rutgers, The State University of New Jersey, Camden, NJ, United States
†Center for Computational and Integrative Biology, Rutgers, The State University of New Jersey, Camden, NJ, United States
[1]Corresponding author: e-mail address: kwangwon.lee@rutgers.edu

Contents

Abstract

Most living organisms on earth experience daily and expected changes from the rotation of the earth. For an organism, the ability to predict and prepare for incoming stresses or resources is a very important skill for survival. This cellular process of measuring daily time of the day is collectively called the circadian clock. Because of its fundamental role in survival in nature, there is a great interest in studying the natural variation of the circadian clock. However, characterizing the genetic and molecular mechanisms underlying natural variation of circadian clocks remains a challenging task. In this chapter, we will summarize the progress in studying natural variation of the circadian clock in the successful eukaryotic model *Neurospora*, which led to discovering many design

Advances in Genetics, Volume 99
ISSN 0065-2660
https://doi.org/10.1016/bs.adgen.2017.09.001

principles of the molecular mechanisms of the eukaryotic circadian clock. Despite the success of the system in revealing the molecular mechanisms of the circadian clock, *Neurospora* has not been utilized to extensively study natural variation. We will review the challenges that hindered the natural variation studies in *Neurospora*, and how they were overcome. We will also review the advantages of *Neurospora* for natural variation studies. Since *Neurospora* is the model fungal species for circadian study, it represents over 5 million species of fungi on earth. These fungi play important roles in ecosystems on earth, and as such *Neurospora* could serve as an important model for understanding the ecological role of natural variation in fungal circadian clocks.

1. INTRODUCTION

Natural variation shapes the diversity of life. In nature, variation is selected by ambient environments, such as altitude, depth, seasons, latitude, and longitude, based on their impacts on fitness (Gaston, 2000). Thus, understanding the underlying principles of these variations and their fitness effect is one of the fundamental quests in biology. One main characteristic of all ambient environments to which most living organisms have adapted is daily cycles. Hot daytime follows cold nighttime, and bright daytime follows dark nighttime. Predicting the incoming environmental change is a crucial trait for an organism to escape or prepare for incoming danger, or to take full advantage of a favorable condition. Anticipating to the environmental change is only possible if one has a cellular device capable of measuring time: a biological clock. When the cellular clock runs with a period of about 1 day, it is called a circadian clock. Scientists tried to understand clock variation under controlled conditions using different circadian models other than *Neurospora*. In the 1980s, Pittendrigh and Takamura showed how latitudinal changes impact circadian parameters (phase, period, and amplitude) using *Drosophila* (Pittendrigh, Kyner, & Takamura, 1991; Pittendrigh & Takamura, 1989). Later, several studies showed that naturally occurring polymorphs in the *Drosophila* period gene (*dper*) impact phenotypic variation in locomotion and temperature compensation (Low, Chen, Yildirim, & Edery, 2012; Low, Lim, Ko, & Edery, 2008; Rosato & Kyriacou, 2011; Sawyer et al., 1997). Additionally, the geographical variation of eclosion rhythms and adult diapause was reported (Lankinen & Forsman, 2006). A recent study has also reported latitude-specific sleep variation in the *Drosophila* natural population (Svetec, Zhao, Saelao, Chiu, & Begun, 2015). Latitude-specific circadian variation has also been observed in other eukaryotes like plants, birds, and mammals. In *Arabidopsis*, natural allelic variations are associated with several circadian phenotypes like hypocotyl growth (Coluccio, Sanchez, Kasulin, Yanovsky, & Botto, 2011; Raschke et al., 2015), flowering time (Lempe et al., 2005; Press, Lanctot, & Queitsch, 2016), periodic leaf movements (Michael et al., 2003; Swarup et al., 1999), and temperature compensation (Edwards, Lynn, Gyula, Nagy, & Millar, 2005). These same allelic variations are also associated with the phenotypes for breeding time (Hau, 2001) and diurnal and seasonal variation of hormone secretion (Dawson, King, Bentley, & Ball, 2001; Reierth, Van't Hof, & Stokkan, 1999) in birds, and seasonal variation of melatonin circadian rhythms (Adamsson, Laike, & Morita, 2016) and latitudinal and seasonal specific qualitative and quantitative variations of sleep (Friborg, Bjorvatn, Amponsah, & Pallesen, 2012) in humans. Based on the population genetic studies, it has also been reported that human clock gene variants are associated with time of day preferences (Allebrandt & Roenneberg, 2008). Genome analysis of human populations at different latitudes revealed a correlation between different day-length (photoperiod) variations of clock genes and certain psychiatric disorders (Forni et al., 2014). Thus, research into natural variation of circadian clocks is no longer limited to model organisms.

In this chapter, we reviewed *Neurospora* as a potential model system for understanding the molecular mechanisms of clock variation and its consequences. Although *Neurospora* is a successful circadian model in eukaryotes (Dunlap & Loros, 2017), the molecular mechanisms of natural variation have not been fully explored. We hope this review will guide researchers to understand the importance of *Neurospora* in investigating the ecological importance of circadian clocks in fungi.

2. *NEUROSPORA* IN NATURE

2.1 The Genus of *Neurospora*

Red bread mold was first reported by Léveillé in France in the 1840s. This fungal species was later incorporated into the genus *Monilia* as *Monilia sitophila*. By the early 20th century, this fungus had been identified across Europe and the United States. Experimental reports showed that this fungus could grow in artificial cultures. In the late 1920s, based on the morphological (conidia and perithecia) similarities, some of the *M. sitophila* group species were transferred into the new genus *Neurospora* and named as *N. sitophila*, *N. crassa*, *N. tetrasperma*, and *N. erythraea* (Shear & Dodge, 1927). Over the next few decades, several fungal species across the globe were studied and classified under the genus *Neurospora*. In the 1970s, Perkins and his colleagues collected over 4500 *Neurospora* isolates from nature, analyzed their physiological characteristics, and classified them into groups: heterothallic (*N. crassa*, *N. discrete*, *N. intermedia*, and *N. sitophila*), pseudohomothallic (*N. tetrasperma*), and homothallic (*N. africana*, *N. dodgei*, *N. galapagosensis*, *N. lineolate*, *N. terricola*, and *N. pannonica*) (Perkins & Turner, 1988). *Neurospora* has three different types of spores: a sexual spore called an ascospore, and two asexual spores (also called conidia) called macroconidia and microconidia (Springer, 1993). Almost all non-homothallic species of *Neurospora* are conidiating species (Turner, Perkins, & Fairfield, 2001). *Neurospora* species are also widely distributed geographically. Initially, *Neurospora* species were found only in tropical and subtropical regions, but were later identified across the globe. In the laboratory, one can produce hybrid populations by crossing *N. crassa* and *N. intermedia*, although the reproductive success rate is lower than that of a cross within species. This hybrid population has been a useful tool in testing ecological factors including reproductive isolation and speciation (Dettman, Anderson, & Kohn, 2008, 2010; Dettman, Jacobson, Turner, Pringle, & Taylor, 2003). Additionally, biological and phylogenetic species of the *Neurospora* genus have been well characterized by Dettman and his colleagues (Dettman, Harbinski, & Taylor, 2001; Dettman, Jacobson, & Taylor, 2003, 2006; Dettman, Jacobson, Turner, et al., 2003), which helps to understand clock variation within different ecotypes.

2.2 Natural Habitats

Ironically, the natural habitats of *Neurospora* are not well understood. Although there is vast amount of knowledge in *Neurospora* biology, most of the studies have been done in a controlled laboratory setting. There are few well-established ecological observations of *Neurospora*'s natural habitat. *Neurospora* is a fire-adapted fungus. The sexual spore is resistant to most environmental stresses and requires a heat shock or a furfural gas that is generated during the fire to germinate (Emerson, 1948; Goddard, 1935). This provides an advantage for local colonization of *Neurospora* after wildfires, as it has been shown that most of the natural strains of *Neurospora* are collected after wildfires (Jacobson et al., 2006). This suggests that *Neurospora* is not a competitive organism in the natural environment unless a condition is right; it cannot colonize on natural substrates unless other natural microbes are absent in a given experimental condition (Lee, 2012). The preferred natural habitat of *Neurospora* has also been shown to vary regionally. In North America, *Neurospora* prefer to colonize beneath tree bark, whereas in Europe they prefer to colonize above tree bark (Jacobson et al., 2006, 2004).

2.3 Collection of *Neurospora* Ecotypes

There are more than 6000 natural ecotypes that have been collected and deposited in the FGSC (Fungal Genetics Stock Center), including an informative record of the collection site of each species deposited. Among the strains deposited in the FGSC, there were unique geographical collection sites for each of the three major *Neurospora* species: *N. intermedia* (253 sites), *N. crassa* (114 sites), and *N. discreta* (85 sites) (Fig. 1). The *N. discreta* species complex is a large clade of species that is different from the clade that includes *N. crassa*. Unlike other species of *Neurospora* whose habitat is mostly in temperate regions (Turner et al., 2001), *N. discreta* has successfully adapted to habitats in diverse latitudinal environments and thus will likely be an ideal species for studying the local adaptation of circadian clocks (Fig. 1).

2.4 *Neurospora* as a Model for Clock Variation Studies

Among all filamentous fungi, *Neurospora* has been widely used to understand several aspects of biochemistry, ecology, evolution, genetics, and circadian biology. *Neurospora* is a haploid filamentous fungus, which has a larger evolutionary timescale compared to other model systems (Baker, Loros, & Dunlap, 2012; Dunlap & Loros, 2017; Turner et al., 2001). *Neurospora* has been well adapted to different latitudes and longitudes across the globe. Therefore, it can potentially serve as a model for understanding several ecological and evolutionary aspects including natural variation. Around the 1970s, Perkins and his colleagues started collecting different species of *Neurospora* across the globe. Most of the collections were submitted to FGSC (Perkins & Turner, 1988; Perkins, Turner, & Barry, 1976; Turner et al., 2001). As mentioned in Section 2.3, there are more than 6000 *Neurospora* ecotypes deposited in FGSC. Among those collected by the FGSC, the most well-characterized reference species is *N. crassa*, the model species that has been widely used to understand the biological aspects of *Neurospora*. The genome of *Neurospora* is relatively small compared to other eukaryotes, and there is a strong synteny of genome among species, especially in *N. crassa*. Additionally, more than 11,000 knockout mutant strains were generated for understanding the gene functions and associated morphology

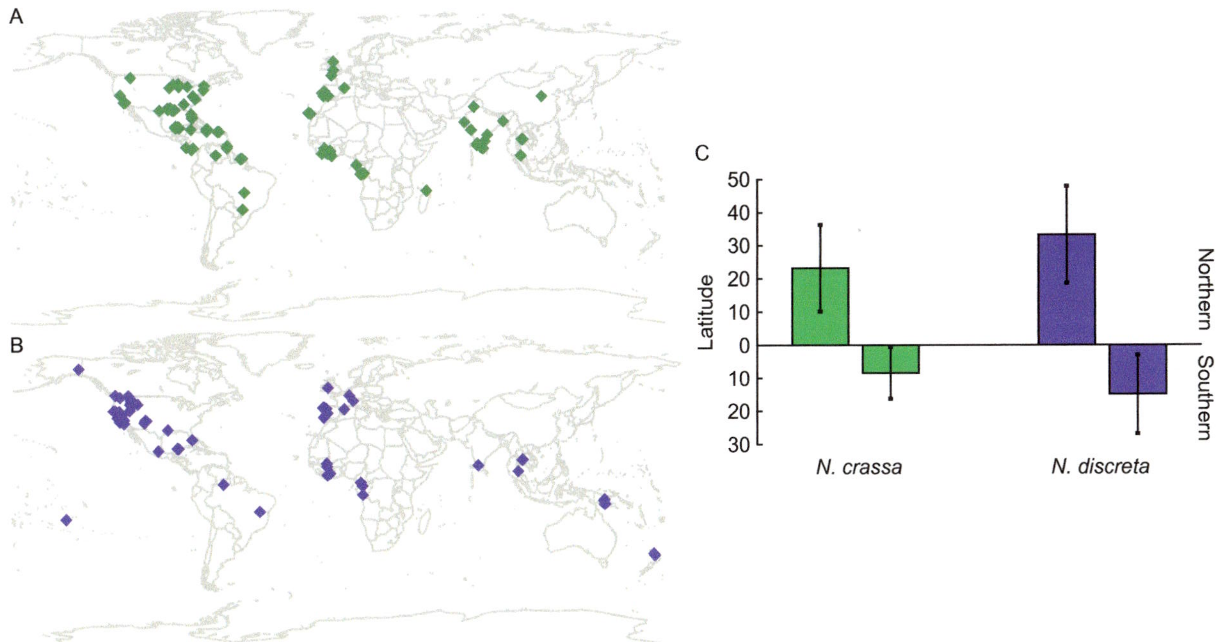

Fig. 1 Distribution of natural ecotypes of *N. crassa* (A) and *N. discreta* (B). Among the natural ecotypes deposited in Fungal Genetics Stock Center, we chose 114 *N. crassa* and 85 *N. discreta* strains that have a complete collection record without redundancy. The distributions of these representative strains are plotted by the collection sites (C). Error bars represent one standard deviation of the mean. The distribution between two species in the Northern hemisphere had a statistically significant difference (t-test, $P < 0.0001$).

(Colot et al., 2006; McCluskey, Wiest, & Plamann, 2010). In the laboratory, the maintenance of *Neurospora* is easy and economical. Since it has a short life cycle and can produce a high number of progeny by both asexual and sexual reproduction, it is a good model system for studying population genetics. The most documented circadian-regulated phenotype in *Neurospora* is asexual development, which is a fitness trait (Brandt, 1953; Pittendrigh, Bruce, Rosensweig, & Rubin, 1959). Based on multiple advantages, *Neurospora* can serve as a potential model to understand natural variations of the circadian clock and its fitness.

3. THE CIRCADIAN CLOCK OF *NEUROSPORA*

Our intention in this section is not an in-depth review of all known molecular mechanisms of the *Neurospora* circadian clock, but rather a brief review to understand studies in natural variation of the *Neurospora* clock. We recommend to readers other excellent reviews on *Neurospora* circadian clocks for a more comprehensive description (Dunlap & Loros, 2017; Heintzen & Liu, 2007; Hurley, Loros, & Dunlap, 2015; Lakin-Thomas, Bell-Pedersen, & Brody, 2011).

3.1 FRQ-Dependent Oscillators

Although *Neurospora* is a simple eukaryotic organism in comparison to higher eukaryotes, the molecular structure of the *Neurospora* circadian clock shows a significant complexity, thus providing an opportunity to dissect network motifs for more complex eukaryotic circadian systems (Bell-Pedersen, Crosthwaite, Lakin-Thomas, Merrow, & Okland, 2001). This molecular structure is composed of the negative components FREQUENCY (FRQ) and VIVID (VVD), and the positive components WHITE COLLAR-1 (WC-1) and WHITE COLLAR-2 (WC-2) (Baker et al., 2012; Cheng, Yang, Gardner, & Liu, 2002; Dunlap & Loros, 2017; Hurley et al., 2015). Traditionally, clock studies were aimed at studying three aspects of the clock: input, the oscillator, and output. In *Neurospora*, FRQ is a key component of the oscillator, WC-1 and WC-2 are part of both the input and the oscillator, and VIVID (VVD) is part of both the input and the output (He & Liu, 2005a; Schneider et al., 2009; Zoltowski & Crane, 2008). It has been shown that WC-1 is a blue light receptor for the circadian oscillator (Froehlich, Liu, Loros, & Dunlap, 2002; He et al., 2002), and FRQ-interacting RNA helicase (FRH) is a protein which interacts with the clock component FRQ (Cheng, He, He, Wang, & Liu, 2005). Several studies have reported the role of casein kinases 1 and 2 (CK1 and CK2) in the negative feedback loop (He et al., 2006; Mehra et al., 2009; Querfurth et al., 2011). The key positive elements in the feedback loop contain PAS domain(s), through which they form homo- or heterodimers. This includes VVD, another PAS protein, and a blue light receptor, which was shown to form a second connected negative feedback loop through WC-1 (Heintzen, Loros, & Dunlap, 2001; Schwerdtfeger & Linden, 2003). The emerging picture is that the clock is composed of multiple regulatory loops.

3.1.1 Simple and Coupled Loops

3.1.1.1 Positive Regulation of WC-1

WC-1 is a blue light receptor and a transcription-binding factor (Froehlich et al., 2002; He et al., 2002) and regulates its own expression (Ballario et al., 1996; Cheng, Yang, Wang, He, & Liu, 2003). When either of the functional WC-1 or WC-2 proteins is absent in a cell, *wc-1* transcription is not light induced (Ballario et al., 1996). Later, it was shown that WC-1 and WC-2 work as a heterodimer activator complex (white collar complex, WCC) (Cheng, He, Yang, Wang, & Liu, 2003; Cheng et al., 2002; Froehlich et al., 2002; He et al., 2002; Talora, Franchi, Linden, Ballario, & Macino, 1999). Although at the surface WCC seems to have a simple genetic regulation with the forward regulation feedback loop, there is more complex regulation at the molecular level.

3.1.1.2 Coupled Loops in WCC

WC-1 negatively regulates WC-2 expression (Cheng, Yang, et al., 2003) and, thus, forms a coupled loop within WCC regulation. WC-1 is a multifunctional protein that contains motifs for sensing the environment and protein–protein interactions, and activation domains with specialized roles in the light and darkness (Cheng, Yang, et al., 2003; Lee, Dunlap, & Loros, 2003; Toyota, Onai, & Nakashima, 2002). Expression of *wc-1* appears to be regulated for its multiple functions, with three promoters identified as necessary for the transcription of *wc-1* (Kaldi, Gonzalez, & Brunner, 2006). One of the promoters, P_{dist}, is responsible for the expression of WC-1 in darkness, and its transcription activity is dependent on WC-1 function. The transcriptional activity of another promoter, P_{prox}, is independent of WC-1 in the dark. The last promoter, P_{int}, is located in the open reading frame of the *wc-1* gene. The transcription activity of the promoter P_{int} is insensitive to a light signal. The alternative translational start site in the P_{int}-originated transcript produces a truncated

WC-1 protein that is lacking the amino-terminal activation domain, NpolyQ, whose activation function influences the circadian period of an organism (Lee et al., 2003; Michael et al., 2007; Toyota et al., 2002). The altered transcriptional activity of *wc-1* also leads to an altered phase of an organism (Kaldi et al., 2006). It appears that WCC can regulate WC-1 expression by choosing different regulatory sites depending on ambient light conditions.

3.1.1.3 Negative Regulation of FRQ

One of the major conceptual breakthroughs in circadian clock studies was the demonstration of the negative feedback loop of the negative regulator, FRQ (Aronson, Johnson, Loros, & Dunlap, 1994). In this classical study, the authors used an overexpression construct of FRQ. When FRQ was overexpressed in a cell by an inducer, its own transcription was downregulated. The detailed mechanisms of the negative feedback mechanisms, however, have been shown to be more than simple negative regulation. Constitutive expression of *qrf* (*frequency antisense transcript*) negatively regulated the level of *frq*, and the absence of *qrf* compromises the light-dependent resetting of clock (Li, Joska, Ruesch, Coster, & Belden, 2015; Xue et al., 2014).

3.1.2 Interlocked Loops in WCC/FRQ

WCC and FRQ control each other's expression through coupled positive and negative regulatory loops.

3.1.2.1 Negative Feedback Loop

The WCC binds to the proximal light regulatory element (PLRE) and clock-responsive element (C-box) in the *frq* promoter. Binding of the WCC to the PLRE leads to acute *frq* induction upon light exposure, which resets the circadian clock. Binding of the WCC to the C-box activates the transcription of *frq*, which is involved in the transcriptional/translational feedback loop of the core circadian oscillator. However, binding of the WCC to the C-box can be interrupted by nucleosomes. SWI–SNF chromatin-remodeling complex clears nucleosomes, promotes the binding of the WCC to the C-box, and induces the expression of *frq* (Dunlap & Loros, 2017; Wang, Kettenbach, Gerber, Loros, & Dunlap, 2014). After the *frq* gene is translated, FRQ homodimers bind to FRH in the cytoplasm and form the FFC (negative elements), which enhances the stability of FRQ (Cheng et al., 2005; Guo, Cheng, & Liu, 2010). Upon localization of FFC into the nucleus, several kinases interact with the FFC complex to phosphorylate FRQ (Baker et al., 2012; Hurley et al., 2015). Of these kinases, casein kinases (CK1a and CK2) are particularly associated with clock-dependent phosphorylation (He et al., 2006; Mehra et al., 2009; Querfurth et al., 2011). CK1a interacts with FFC through the FRQ/CK1a interaction domain (FCD), which is located near the three phosphorylation sites of the *frq* gene. The interaction of CK1a with the FFC plays a crucial role in the negative feedback mechanism. CK1a also phosphorylates and inactivates the WCC. This process is mediated by the physical interaction between the FFC and the WCC (He et al., 2006). As the activator function of the WCC diminishes, the level of *frq* transcription also diminishes. After the rate of *frq* transcription and translation slows down, the level of FRQ in the nucleus diminishes (Aronson et al., 1994; He, Cheng, Yang, Yu, & Liu, 2003). The low level of FRQ in the nucleus allows WCC in the nucleus to resume transcription of *frq*, which closes the negative feedback loop. The period of circadian oscillation of the clock gene FRQ is about 22 h, which includes the time for synthesis, phosphorylation, and degradation of the FRQ protein (Dunlap & Loros, 2017; Hurley et al., 2015; Merrow, Garceau, & Dunlap, 1997).

3.1.2.2 Positive Feedback Loop

FRQ is involved not only in negative regulation of its activators (WCC) but also in the positive regulation of WCC. FRQ is involved in the upregulation of WC-1 at the posttranscriptional level and in the upregulation of WC-2 at the transcriptional level (Cheng, Yang, & Liu, 2001; Lee, Loros, & Dunlap, 2000). Thus, FRQ plays a role in coupling the negative and positive loops. This interconnected loop has also been reported in *Drosophila* and mammals (Glossop, Lyons, & Hardin, 1999; Shearman et al., 2000). In all cases, the interconnected loop plays a crucial role in maintaining the robustness of the circadian oscillation (Cheng et al., 2001).

3.1.3 Posttranslational Modification of WCC and FFC

In addition to transcriptional/translational feedback regulation, posttranslational regulation plays a role in circadian regulation (Brunner & Schafmeier, 2006). The phosphorylation status of the WCC is important for its activation and functionality. Phosphorylation of the WCC is FRQ dependent, and hyperphosphorylated WCC is an inactive form of WCC (He & Liu, 2005b; Schafmeier et al., 2005). Upon exposure to light, the WCC forms a complex that is much larger than the WCC in darkness (Froehlich et al., 2002; He & Liu, 2005b). Because the number of FFCs in a cell is lower than the number of WCCs, it was speculated that the negative function of FRQ–FRH could be achieved by recruiting kinases for phosphorylation of the WCC. There are several kinases that are known to phosphorylate clock proteins (Cheng et al., 2005; Froehlich, Noh, Vierstra, Loros, & Dunlap, 2005; Görl et al., 2001; He et al., 2006), including casein kinases, which phosphorylate the WCC and FRQ. To finish the negative loop, the negative elements are degraded. This is achieved through the ubiquitin–proteasome system much like in mammals and *Drosophila* (He & Liu, 2005a). There are four kinases and two phosphatases involved in FRQ phosphorylation (Garceau, Liu, Loros, & Dunlap, 1997; Görl et al., 2001; Liu, Loros, & Dunlap, 2000; Pregueiro, Liu, Baker, Dunlap, & Loros, 2006; Yang, Cheng, & Liu, 2002; Yang, Cheng, Zhi, & Liu, 2001; Yang et al., 2004). Eventually, hyperphosphorylated FRQ is recruited by FWD-1, an F-box/WD-40 repeat-containing protein. FWD-1, a substrate-recruiting subunit of an SCF (SKP/Cullin/F-box)-type ubiquitin ligase, physically interacts with hyperphosphorylated FRQ, which leads to the degradation of the FRQ protein (He et al., 2003; He & Liu, 2005a). For the simplicity of the model, we did not include the degradation of FRQ by ubiquitination in our model. In the absence of FRQ, FRH can interact with WCC (Guo et al., 2010); however, the role of FRH in the circadian clock was unclear. It has been shown that FRH also plays a vital role in cell viability (Hurley, Larrondo, Loros, & Dunlap, 2013; Lauinger, Diernfellner, Falk, & Brunner, 2014; Shi, Collett, Loros, & Dunlap, 2010), though a recent study has shown that FRH may not play an essential role in the circadian clock (Conrad et al., 2016).

3.1.4 Negative Feedback Loop in WCC/FRQ/VVD

The gene VVD also forms negative feedback loops. VVD is another blue light receptor that has two major roles: sensing the change of light intensities and photoadaptation (He & Liu, 2005b; Heintzen et al., 2001; Schwerdtfeger & Linden, 2003; Shrode, Lewis, White, Bell-Pedersen, & Ebbole, 2001). The role of VVD in endogenous clock regulation is that of a clock-controlled output gene, and also a modulator of the light input pathway to the clock (Chen, DeMay, Gladfelter, Dunlap, & Loros, 2010; Heintzen et al., 2001; Hunt, Thompson, Elvin, & Heintzen, 2010). VVD is one of the early light-response genes that are activated by WCC (He & Liu, 2005b; Heintzen et al., 2001; Schwerdtfeger & Linden, 2003). Once VVD is expressed, it negatively regulates its own expression by blocking the activator function of the WCC by directly binding to the WCC

(Chen et al., 2010; Hunt et al., 2010). This photoadaptation function ensures the circadian oscillator runs during the daytime (Elvin, Loros, Dunlap, & Heintzen, 2005). VVD also plays a key role in resetting circadian rhythm at dusk (light to dark transition) by enhancing *frq* degradation, and thus, de novo synthesis occurs, causing a new cycle to begin (Elvin et al., 2005).

3.2 FRQ-Independent Oscillators

In the 1980s, Feldman's lab first reported FLO (FRQ less oscillator) rhythms using a *frq* mutant (Loros, Richman, & Feldman, 1986). Over the years, several studies were done to understand the mechanism underlying the developmental rhythms by FLO. In the late 1990s, a report was published that frq^{9} mutants can be entrained by temperature, but not with light (Chang & Nakashima, 1997; Merrow, Brunner, & Roenneberg, 1999). Furthermore, Lakin-Thomas and Brody published that deficiency of lipids can restore the developmental rhythms in the absence of a core circadian oscillator, which includes photoreceptors like *wc-1* and *wc-2* (Lakin-Thomas & Brody, 2000). These observations support that temperature and metabolism can play an essential role in generating developmental rhythms in addition to the core circadian oscillator. Based on these studies, a two-oscillator model has been proposed; the core circadian oscillator responds to light and FLO responds to nonphotic signals such as temperature. These two oscillators may or may not be coupled to generate developmental rhythms (Bell-Pedersen et al., 2001). In addition to a Choline-Deficient Oscillator, CDO (Lakin-Thomas, 1998; Lakin-Thomas & Brody, 2000), a Farnesol- or Geraniol-based Oscillator, FGO (Granshaw, Tsukamoto, & Brody, 2003), and a Cryptochrome-dependent Oscillator, CRO (Nsa et al., 2015), have been identified in *N. crassa*. Additionally, microarray studies have shown that several clock-controlled genes are rhythmic in the absence of *frq* (Correa et al., 2003). This suggests that different FLOs can take over the developmental rhythms. The relationships between the FRQ-based oscillator and FLOs have yet to be characterized.

In the context of natural variation, metabolic enzymes and metabolic components, e.g., mitochondria, have been reported (Table 1). Interestingly, allelic variations of *nit-2* for nitrate reductase and *nit-4* for nitrogen regulation have been observed in *N. sitophila* and *N. intermedia*, respectively. Recently, a study has been done on the *frq*-less Nitrate Reductase Oscillator in *N. crassa*, NRO (Christensen et al., 2004), in which NIT-2 and NIT4 are the positive elements of the feedback loop.

Table 1 Natural Variant Studies in *Neurospora*

Species	Category	Traits	References
N. crassa	Chromosomal modifications	Satellite	Perkins, Metzenberg, Raju, Selker, and Barry (1986)
		Tandem duplications	Selker and Stevens (1987)
		Translocations	Perkins and Turner (1988)
	Enzymes	β-Glucosidase	Mahadevan and Eberhart (1964)
		Cellulase	Eberhart, Beck, and Goolsby (1977)
		Fatty acid synthesis	Goodrich-Tanrikulu, Stafford, and Jacobson (1999)
		Invertase	Yu, Garrett, and Sussman (1971)
		Leucyl-tRNA synthetase	Beauchamp, Horn, and Gross (1977)
		Nitrogen regulation	Young and Marzluf (1991)
		Ornithine transcarbamylase	Grindle and Davis (1970)
		Protease	Hanson and Marzluf (1975)
		Tyrosinase	Horowitz, Fling, Macleod, and Sueoka (1961); Ruegg, Ammer, and Lerch (1982)
	Cell cycle	Meiosis impaired	Smith (1975)
		Recombination repression	Catcheside (1975)
		Sexual phase recessives	Leslie and Raju (1985); Raju and Leslie (1992)
		Synaptic sequence	Catcheside (1981)

Continued

Table 1 Natural Variant Studies in *Neurospora*—cont'd

Species	Category	Traits	References
	Mitochondria	Mitochondrial DNA variants	Collins and Lambowitz (1983); Mannella, Pittenger, and Lambowitz (1979); Taylor, Smolich, and May (1986)
		Mitochondrial plasmids	Griffiths (1995); Taylor, Smolich, and May (1985); Taylor et al. (1986)
		Senescence	Navaraj, Pandit, and Maheshwari (2000)
	Nutrients		Leslie and Raju (1985); Perkins et al. (1976)
	Reproductive success		Dettman, Jacobson, Turner, et al. (2003)
	Resistance to *Sk*		Campbell and Turner (1987)
	Resistance to toxic agents	Canavanine	Perkins, Radford, Newmeyer, and Bjorkman (1982)
		Cadmium resistance	Levine and Marzluf (1989)
		Surfactant resistant	Al-Saqur and Smith (1980)
		Triphenyl tetrazolium chloride	Perkins et al. (1982)
	Vegetative incompatibility	*het* loci	Mylyk (1975, 1976); Perkins (1975)
		Suppressor	Newmeyer (1970)
N. discreta	Enzymes	Fatty acid synthesis	Goodrich-Tanrikulu et al. (1999)
	Mitochondria	Mitochondrial plasmids	He, Nastasja de, Bok, and Griffiths (2000)
	Speciation	Phylogenetic variation	Dettman et al. (2006)

N. intermedia	Chromosomal modifications	Tandem duplication	Selker and Stevens (1987)
		Translocations	Perkins and Turner (1988)
	Enzymes	Allozymes	Spieth (1975)
		β-Glucosidases	Perkins and Turner (1988)
		Cellulase	Perkins and Turner (1988)
		Estreases	Reddy and Threlkeld (1972)
		Nitrate reductase	Blakely and Srb (1962)
		Tyrosinase	Perkins and Turner (1988)
		Microcycle conidiation	Turner et al. (2001)
		Nitrogen regulation	Young and Marzluf (1991)
		Fatty acid synthesis	Goodrich-Tanrikulu et al. (1999)
	Mitochondria	Mitochondrial DNA variants	Collins and Lambowitz (1983)
		Mitochondrial plasmids	Griffiths (1995); Natvig, May, and Taylor (1984)
		Senescence	Bertrand, Chan, and Griffiths (1985); Bertrand, Griffiths, Court, and Cheng (1986); Griffiths and Bertrand (1984); Rieck, Griffiths, and Bertrand (1982)
	Resistance to toxic agents	Surfactant resistant	Al-Saqur and Smith (1980)
	Reproductive success		Dettman, Jacobson, Turner, et al. (2003)
	Spore killer	Resistance to *Sk*	Campbell and Turner (1987)
	Virus-like particles		Perkins and Turner (1988); Tuveson and Peterson (1972)

Continued

Table 1 Natural Variant Studies in *Neurospora*—cont'd

Species	Category	Traits	References
N. sitophilia	Chromosomal modifications	Tandem duplication	Selker and Stevens (1987)
	Enzymes	Amylases	Reddy and Threlkeld (1971a, 1971b, 1972)
		Nitrogen regulation	Grove and Marzluf (1981); Perkins and Turner (1988); Young and Marzluf (1991)
		Fatty acid synthesis	Goodrich-Tanrikulu et al. (1999)
		Tyrosinase	Perkins and Turner (1988)
	Mitochondria	Mitochondrial plasmids	Griffiths (1995)
		Mitochondrial DNA variants	Collins and Lambowitz (1983)
N. tetrasperma	Chromosomal modifications	Tandem duplication	Selker and Stevens (1987)
	Enzymes	Aryl-sulfatase	Metzenberg and Ahlgren (1971)
		Estreases	Reddy and Threlkeld (1971a)
		Fatty acid synthesis	Goodrich-Tanrikulu et al. (1999)
	Mitochondria	Mitochondrial plasmids	Griffiths (1995); Natvig et al. (1984)

4. NATURAL VARIATION AND the CIRCADIAN CLOCK

There is a large volume of literature on natural variation studies using *Neurospora*. In Section 4.1, we will highlight a few selected studies that are worth revisiting in the study of natural variation of the *Neurospora* circadian clock. Considering the recent advancements in our understanding of the existence of metabolic oscillations and their cross talk with the genetic oscillator, we believe that old literature on natural variation studies in metabolites and enzymes can provide potentially new insights for studying natural variation in *Neurospora* circadian clocks.

4.1 Natural Variation in *Neurospora* Ecotypes

From the 1920s, *Neurospora* wild isolates from different natural habitats were grown in synthetic media to study physiological variation of ascospores, conidia, and perithecia (female sexual organ). The ability to grow *Neurospora* in a defined artificial media made *Neurospora* an attractive model organism. Different species of *Neurospora* have been characterized based on physiological variations such as number of asci, ascospore size, pigmentation, the size of perithecia, and mating behavior (Perkins et al., 1976; Shear & Dodge, 1927; Tai, 1935). Several early studies have shown phenotypic variation in different environmental conditions: speed of conidial germination was influenced by the ambient temperature (Shear & Dodge, 1927; Thom & Ayers, 1916), a higher ascus abortion rate in the Texas strain of *N. tetrasperma* has been reported on different media (Dodge, 1939), and mycelial growth and sexual development of *N. lineolate* varied in different media conditions (Frederick, Uecker, & Benjamin, 1969). Other investigations on nutritional factors suggest that variations in perithecia formation occur due to the concentrations of glucose and potassium nitrate of a synthetic medium, which favors sexual reproduction in *Neurospora* (Westergaard & Mitchell, 1947). Around the 1950s, scientists started to understand the biochemistry of conidia and mycelia using *Neurospora*. A 1954 study reported the importance of light and oxygen in the synthesis of carotenoids of *N. crassa* (Zalokar, 1954). Furthermore, variations in respiratory enzymes such as NAD nucleotidase, β-galactosidase, and β-glucosidase have been reported in conidia compared to the mycelia (Eberhart, 1961; Zalokar, 1959; Zalokar & Cochrane, 1956). The activity of the citric acid cycle was shown to decrease in conidia. Therefore, Weiss and his colleague to begin to understand the variation in activity of mitochondrial enzymes such as oxidative and glycolytic enzymes. Among them, variation of glycolytic enzymes (ethanol dehydrogenase and pyruvate carboxylase) has been reported in conidial and mycelial cultures (Weiss & Turian, 1966).

Several studies have also investigated the genetic control associated with phenotypic variations of enzymes in wild isolates of *Neurospora*: *ars* for arylsulfatase (Metzenberg & Ahlgren, 1971); *arg-12* for ornithine transcarbamylase (Grindle & Davis, 1970); *leu-5* for leucyl-tRNA synthetase (Beauchamp et al., 1977); *inv* for invertase (Yu et al., 1971); *nit-4* for nitrate reductase (Perkins & Turner, 1988); *pts-1* for protease (Hanson & Marzluf, 1975); *nit-2*, *nmr* for nitrogen regulation (Grove & Marzluf, 1981; Young & Marzluf, 1991); and *T* locus for tyrosinase (Horowitz et al., 1961; Ruegg et al., 1982). Several other studies have investigated the variation in other enzymes such as amylases, allozymes, β-glucosidases, cellulases, and esterases (Perkins & Turner, 1988; Perkins et al., 1976; Reddy & Threlkeld, 1971a, 1971b, 1972; Spieth, 1975; Turner et al., 2001). Interestingly, most of the enzymatic variations have been associated with enzymes involved in metabolic pathways. In addition to enzymatic variations, mitochondrial variants have been reported in natural populations of *Neurospora*.

In the late 1970s, mitochondrial DNA variants were reported in *N. crassa* (Mannella et al., 1979). Later, mitochondrial DNA variants were identified in inter and intraspecific populations of *Neurospora* (Collins & Lambowitz, 1983; Griffiths, 1995; Perkins & Turner, 1988; Taylor, 1986; Taylor, Koupel, & Whitmer, 1987; Turner et al., 2001). Additionally, natural variation in the shape and size of mitochondrial plasmids was found in different species of *Neurospora*. Integration of circular plasmids into the mitochondrial genome can interfere with mitochondrial functions, subsequently causing death of cultures (Senescence), which has been observed mainly in *N. intermedia* (Griffiths, 1992, 1995; Maheshwari & Navaraj, 2008; Perkins & Turner, 1988; Turner et al., 2001). Table 1 summarizes more natural variation studies, which might be useful for characterizing the natural variation in circadian clock. For more information on natural variations and *Neurospora* natural population, there are excellent reviews available (Perkins & Turner, 1988; Perkins et al., 1976; Turner et al., 2001).

4.2 Natural Variation of Core Clock Genes

Using inverted race tube assays (Park & Lee, 2004), 143 ecotypes of *N. crassa* have been characterized. There was a significant variation in the circadian period, phase, and temperature compensation among the ecotypes (Michael et al., 2007). In addition to the phenotypic variation of the clock, genetic variation has been studied in four key clock genes, *frq*, *wc-1*, *wc-2*, and *vvd*. A subsequent study tested the hypothesis that the copy number variation of Simple Sequence Repeats (SSRs) in the clock genes could serve as a way of creating phenotypic variation in *Neurospora*. This hypothesis was based on one of the conclusions from a parallel genome-wide study of the SSR distribution in *Neurospora* genomes that trinucleotide SSRs in the exon might play a role in the variation of protein functions in *N. crassa* (Kim et al., 2008). All core clock genes contain SSR repeats. Most interestingly, a photoreceptor and transcription factor WC-1 has two large SSR repeats: one in the amino-terminal, NpolyQ, and one in the carboxyl-terminal, CpolyQH. In the *wc-1* alleles that lack NpolyQ, *wc-1*^{MK1}, and *wc-1*$^{rhy-2}$, there is no functional clock running, whereas *frq* is light induced (Cheng, Yang, et al., 2003; Lee et al., 2003; Toyota et al., 2002). This suggests that WC-1 activation domains may function in a modular fashion, and NpolyQ could be a main activation domain for the clock function. The estimated repeat size of NpolyQ was between 17 and 80 base pairs among ecotypes, and the laboratory strain 87-3 had 39 repeats. There was a significant correlation between the size of NpolyQ repeats and the circadian period among 143 natural ecotypes. To test if the size of NpolyQ repeats explains the period variation, two ecotypes with similar structural features except the size of NpolyQ repeats were crossed: FGSC#3223 (29 repeats) and FGSC#4724 (57 repeats). The period of F1 progeny segregated with the size of NpolyQ alleles. The authors concluded that the natural variation in WC-1 NpolyQ of ecotypes may have served for fine tuning their circadian clock to a local habitat (Michael et al., 2007). This suggests that SSR variation could be the answer to how new gene function could arise in a genome with several strong genome defense mechanisms in which any duplicated genome sequences more than 500 nt would be mutated (Aramayo & Selker, 2013).

The conservation of FRQ function in the family Sordariomycetes has been demonstrated by complementing the *Neurospora frq* mutant by expressing the FRQ homolog from *Sordaria fimicola*. The resulting fusion protein had 11 amino acids from *N. crassa* in the amino-terminal with the rest of protein sequence from *S. fimicola* (Merrow & Dunlap, 1994). Encouraged by this success, the sequence of the *frq* gene has been used to establish a phylogenetic relationship among species in the genus of *Neurospora* and in

Ascomycete fungi with a limited success (Lewis & Feldman, 1996; Skupski, Jackson, & Natvig, 1997). A more comprehensive sequence analysis has been done with 5 clock genes (*frq*, *wc-1*, *wc-2*, *frh*, *fwd-1*) in 64 sequenced fungal genomes (Salichos & Rokas, 2010). FRH and FWD-1 appeared to be the most ancient proteins, followed by WC-1 and WC-2. Interestingly, FRQ appeared to be the most recent protein and appears in only three families: Sordariomycetes, Leotiomycetes, and Dothideomycetes in the Ascomycetes (Salichos & Rokas, 2010).

4.3 Technical Challenges for Studying Clock Variation in *Neurospora* Natural Populations

As is true for other biological research, having an easily assayable phenotype is important for a successful outcome. For *Neurospora* circadian studies, the rhythmic asexual development (conidiation) was a blessing. In the 1950s, the rhythmic pattern of conidiation in *Neurospora* was reported (Brandt, 1953; Pittendrigh et al., 1959). Unlike the unicellular yeast, filamentous fungi like *Neurospora* will grow continuously away from the center of inoculation. Spending the majority of time in the vegetative stage, *Neurospora* will grow as mycelia (transparent or white color). Once approximately 24 h has passed, *Neurospora* will stop growing as mycelia and start producing conidia (orange color). The conidia are produced from a specialized aerial hyphae cell type called a conidiophore (Springer & Yanofsky, 1989). In a macroscopic view, the area of *Neurospora* that produces conidia is seen as denser in comparison to that of mycelia. This dense area is called a "band." When *Neurospora* is grown in a glass tube with a medium on the bottom (race tube), the developmental pattern can be easily visualized. To measure the circadian period (in the constant darkness condition) or phase (in a cyclic light/dark condition), the growing front of the mycelia is marked on the glass tube under the red light, which does not interfere the *Neurospora* circadian rhythm. Period is measured by the distance from one reference point to the next. Phase is measured as the distance from a reference point of environmental cycle, e.g., dark to light transition, or to a reference point of a development. Traditionally, the reference point has been the peak of a conidiation band. Recently, the photographic image is read by software such as Chrono, developed by Roenneberg, to calculate the clock phenotypes of the sample using different reference points, e.g., trough, onset, or offset (Roenneberg & Taylor, 2000). However, visualizing the developmental bands in the natural strains was not a trivial matter. Conidiation is repressed by the accumulated CO_2 produced by the respiration of the cells, which is trapped inside of the race tube (Sargent & Kaltenborn, 1972). Therefore, virtually all circadian studies in *Neurospora* have been done in the band (*bd*) mutant background to visualize the banding phenotype (Sargent & Woodward, 1969). The *bd* mutation enabled the fungus to overcome the adverse effect of CO_2 accumulation on conidiation. Four decades after the discovery of the band mutation, it was shown to be located in the *ras-1* gene, which is involved in producing reactive oxygen species known to stimulate conidiation (Belden et al., 2007). Unfortunately, one cannot use this mutant background to study circadian rhythms of ecotypes. To create the *bd* mutant background, one has to cross an ecotype with a *bd* mutant. In this process, the unique genotype adapted to one local habitat will be lost. The alternative methods to study the ecotypes without crossing them to a *bd* mutant strain were proposed, such as blowing fresh air to remove CO_2 or adding rubidium chloride to the medium (Morgan & Feldman, 1998). These methods were not ideal, however, as blowing air into a race tube tends to spread the hydrophobic conidia into other parts of the media, and adding RuCl causes additional variables in an asexual development of wild ecotypes. One successful assay method to visualize the overt developmental rhythm of natural ecotypes was developed by simply inverting the conventional race tubes. This is known as an inverted race tube assay (Park & Lee, 2004). This modified race tube assay allowed the accumulated CO_2 in the race tube to be ventilated out without blowing fresh air into the race tube, as CO_2 is heavier than air. Because the natural strains grows faster than the band mutants, as the *bd* mutation reduces growth rate, it was necessary for the modified race tube to be longer than the conventional race tube designed for the *bd* mutants (Park & Lee, 2004).

4.4 Methods for Clock Variation Studies

4.4.1 QTL Analysis on Clock Phenotypes

To elucidate the mysteries of gene function caused by genetic variations in the natural population, quantitative genetic techniques have been successfully utilized over the past decades (Alonso-Blanco & Koornneef, 2000). These methods describe how known mutant loci can interact with one another, as well as to isolate new loci in the same pathway (Shimomura et al., 2001). Over the past century, the field of quantitative genetics has grown out of the desire to exploit natural variation and to describe traits with a continuous distribution of phenotypes. An extension of Mendelian principles of polygenic trait is quantitative genetics (phenotypes encoded by multiple loci). Most of the phenotypic variation identified in natural populations is due to quantitative trait loci (QTL), in which multiple loci contribute small effects on phenotypes, as opposed to a single major-effect locus (Alonso-Blanco & Koornneef, 2000).

Haploid organisms have multiple advantages for studying quantitative traits: (1) for assays, a large number of individuals can be maintained clonally, which may reduce the error from environment; (2) they have a single allele at any given locus, so dominance does not contribute to the genetic variation; (3) QTL can be detected after the first generation; and, as in the case of *Neurospora*, (4) they are fast and easy to work with. For these reasons, *Neurospora* can serve as a potential model to elucidate fundamental questions of quantitative genetics for complex behaviors including circadian clocks. As our understanding of the molecular structure of circadian clocks advances, it becomes apparent that the circadian clock is more tightly linked with other cellular machinery than we imagined (Bell-Pedersen et al., 2001). Based on this, one cannot dissociate the circadian clock from the basic cellular metabolism. In the previous efforts to find clock genes, all morphological mutants and any mutants with abnormal metabolisms other than the period phenotypes were excluded from analysis. As we have learned more about the importance of the circadian clock in cellular activities, we have realized that there could be clock genes involved in these basic metabolic activities as well. These genetic loci with subtle clock phenotypes or with essential cellular functions could not be studied using traditional genetic screening methods. QTL analysis could reveal novel clock genes and clock mechanisms which would be overlooked by the traditional forward genetic approaches. QTL analysis is a systematic way of mapping all the multiple genetic loci for subtle quantitative traits and essential cellular functions.

To study the natural allelic variants, three mapping populations were prepared using six *N. crassa* ecotypes from distant geographic locations (N2, FGSC3223 from USA × FGSC4724 Malaysia; N4, FGSC4720 from India × FGSC4715 Haiti; N6, FGSC4825 Ivory Coast × FGSC2223 USA). QTL analysis on the circadian period and entrained phase phenotype identified 43 QTL (Kim, Logsdon, Park, Mezey, & Lee, 2007). Thirteen of 43 QTL colocalized to known clock genes, which supports that QTL analyses can identify clock genes. More surprising was that 30 of 43 QTL (70%) did not colocalize to known clock genes. This suggests that there are many uncharacterized genetic elements whose variation contributed to the variation of the circadian clock for a specific habitat. Identifying QTL (a chromosomal region which could contain multiple genetic elements) is relatively easy in comparison to identifying causative genes and the nature of variation that caused the phenotypic variation. Since *Neurospora* has a relatively small haploid genome (about 40 Mb), it is a promising system to characterize the causative natural allelic variation for clock variation.

4.4.2 Circadian Clocks in Natural Substrates and Natural Environment

Unlike animals, a filamentous fungus cannot move to other locations when one place does not provide a favorable environment. However, a filamentous fungus can grow indefinitely to a new place that might provide a better living condition. This is why filamentous fungi are the largest living organisms on earth. In one recorded case, one genetically identical fungus *Armillaria bulbosa* grew over 1500 years covering over 15 ha with an estimated weigh over 10 tons (Smith, Bruhn, & Anderson, 1992). One might have a hard time justifying ecological studies on circadian rhythms of *A. bulbosa* in a 27-in. race tube under the laboratory conditions. For the reasons that we argued throughout this review, *Neurospora* is the best model system for ecological and evolutionary studies of circadian rhythms. The current challenge is developing an experimental system that utilizes natural substrates in a natural environment, and generates reproducible experimental data for characterizing circadian rhythms of a fungal strain. As the first step toward this goal, *Neurospora* has been grown on different natural substrata (Lee, 2012). Utilizing new functional genomics tools, *Neurospora* could serve as a model system for addressing circadian clock-associated ecological questions (Ellison et al., 2011; Lee, 2012; Lee & Dighton, 2013; Romero-Olivares, Taylor, & Treseder, 2015).

5. FUTURE CHALLENGES

Considering the important ecological role of the fungal kingdom in the ecosystem on earth, we have no doubt that understanding the natural variation of the fungal circadian clock is both interesting and an important area of research. As discussed earlier, we are hopeful that more researchers will invest time in the study of natural variation of fungal circadian clocks due to the many advantages that the system offers. However, we also expect a few tough challenges that require much deliberation and a consensus from the research community. These challenges are present not only in the study of fungal clocks but also in natural variation studies in general.

There is a great advantage in working with a model organism like *Neurospora*, especially when it is a reference strain with a great deal of resources. Because of the large amount of resources invested on a few select model organisms, we now enjoy a detailed understanding of clock mechanisms. In choosing a model reference organism, it was necessary that the three minimum characteristics required to be called an "authentic" circadian clock be present: an approximate 24-h period in constant conditions, the oscillator is entrainable by the local environment, and period is temperature compensated within a physiological range of temperatures. However, as we study the ecotypes (or accessions) of the model organisms that are adapted to a diverse habitat, we find significant deviations in these clocks from that of the reference strain. Circadian clocks have been observed in fungi other than *N. crassa*; however, these clocks are not satisfying the community-agreed criteria to be called an authentic circadian clock (Austin, 1968; Bluhm, Burnham, & Dunkle, 2010; Greene, Keller, Haas, & Bell-Pedersen, 2003; Hevia, Canessa, & Larrondo, 2016; Hevia, Canessa, Muller-Esparza, & Larrondo, 2015; Oliveira et al., 2015; Traeger & Nowrousian, 2015). We believe that it is more interesting to find different kinds of clocks with different properties, which reflect the life history of the organism (or species) and speak volumes about the real biology of the clock in nature. As we classify circadian clocks into "authentic" and "not-so-authentic," we may somehow be deemphasizing the studies of those clocks that are not satisfying the minimum criteria of the authentic clock.

Most of the known clock properties in a model organism have been characterized in a controlled laboratory environment using laboratory strains, which served its purpose well to generate reproducible data. However, to study the ecological roles of the circadian clocks, we need to perform experiments in their natural habitat. In a pleasant spring day, the surface temperature of the ground (where many fungi live) could fluctuate from 0°C to 40°C in a day. The physiological temperature range that we know for temperature compensation spans 20–30°C for a typical laboratory *Neurospora* strain. We envision a need to develop criteria to characterize nonmodel fungal clocks (or ecotypes of model species) in natural habitats.

Several studies on eukaryotes have shown the role of circadian clock in maintaining fitness (Beaver et al., 2002; Gill & Panda, 2011; Green, Tingay, Wang, & Tobin, 2002; Miller et al., 2004; Xu, DiAngelo, Hughes, Hogenesch, & Sehgal, 2011). Even though asexual reproduction is a fitness trait of filamentous fungi like *Neurospora*, natural variations of the circadian clock and associated fitness trait are not well understood. There are different ways of measuring "fitness" in filamentous fungi, but there is no clear consensus in the fungal community. One could measure the growth rate, estimating the amount of nuclei by PCR, number of conidia, germination rate of conidia, number of perithecia (fertilized female organ), number of ascospores (sexual spore), germination of ascospores, etc. (Pringle & Taylor, 2002). According to a classical definition of fitness, one might measure the number of progeny (conidia or ascospore). However, there are many aconidial species that do not produce conidia. Even in conidial species with a known sexual cycle, the number of progeny is greatly changed depending on the media and environment. Many fungal species also have different developmental pathways. For example, *Neurospora* has two different asexual progeny (macroconidia and microconidia) and one sexual progeny (ascospore). Macroconidia contain three to five nuclei, and microconidia contain one nucleus. Depending on the natural habitat a *Neurospora* strain lands on, it will produce different ratios of macroconidia to microconidia (Lee, 2012). It is a challenging task to measure fitness in an unbiased way. We hope that it will prove possible to develop a consensus on what is an appropriate measure of fitness.

In an ideal situation, one would like to study a natural strain in a natural habitat. Unlike animals or plants, it is difficult to study fungi in their natural environment. A luciferase reporter is a great way of monitoring the molecular clock, assuming that the fungus of interest can be transformed with the reporter construct and have its expression measured by a luminometer. One possibility to remedy this difficulty is to create a laboratory setting that closely mimics the natural environment.

6. CONCLUDING REMARKS

Even though *Neurospora* is a classical model for studying the circadian clock, it has not been widely used for understanding the ecological and evolutionary aspects of clocks. In this chapter, we reviewed the challenges, advantages, and promising future of the system to study natural variation of circadian clocks. We proposed that *N. discreta* should be used for studying effects on fitness, as it has greater adaptability to different latitudinal environments compared to other *Neurospora* species (Fig. 1). Since asexual development is one of the most well documented clock-regulated phenotypes, *Neurospora* is an excellent system to study the evolutionary and ecological roles of natural variation of the circadian clock (Fig. 2).

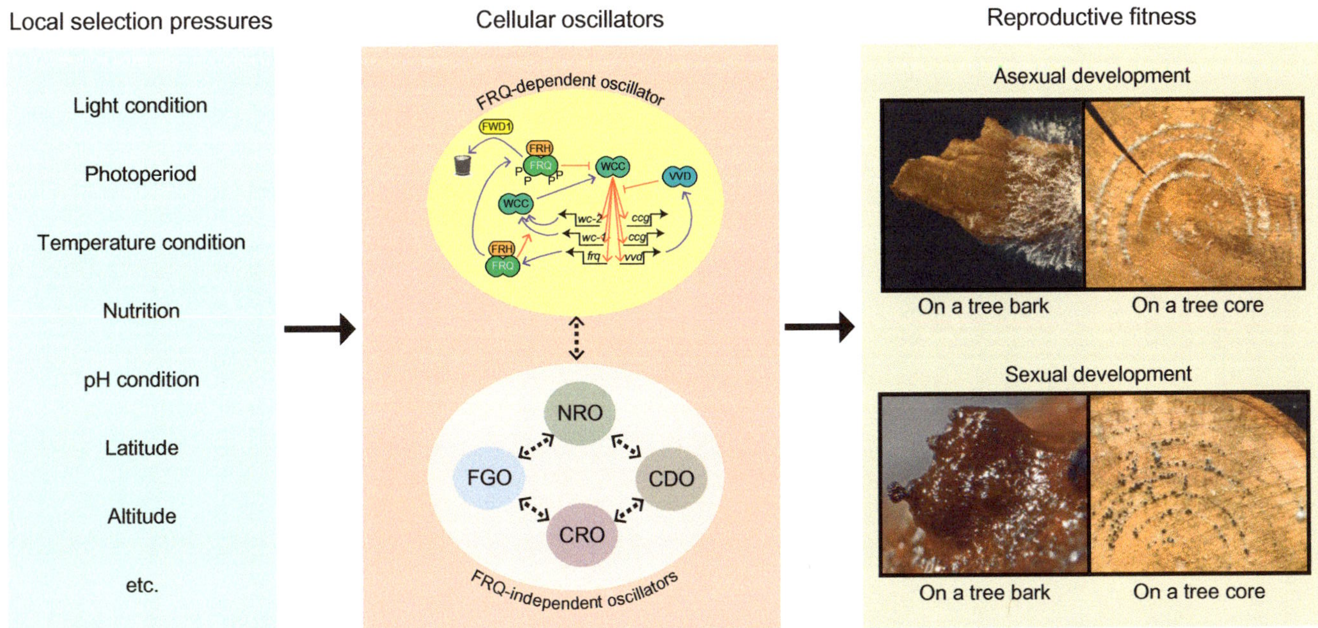

Fig. 2 Overview of natural variation studies in fungal circadian clocks. Each natural strain has been successfully adapted to the local environment. Many layers of ambient environmental factors will function as selection forces. Thus, the variation observed in the circadian rhythm reflects the natural history of the organism. Among the circadian controlled phenotypes, asexual and sexual development in *Neurospora* have been extensively studied. Cellular oscillators: *blue lines* for expressed gene products, *red lines* for upregulation (*arrow*) or downregulation (*line*) of a gene, and *dotted line* for unknown relationships. *Images for a conidial development are modified from Lee, K., & Dighton, J. (2010).* Neurospora*, a potential fungal organism for experimental and evolutionary ecology.* Fungal Biology Reviews, *24(3–4), 85–89. doi:10.1016/j.fbr.2010.09.001 and from Lee, K. (2012). Asexual and sexual developments of* Neurospora crassa *on natural substrata.* Fungal Ecology, *5(2), 223–229. doi:10.1016/j.funeco.2011.09.001. Images of sexual development are modified from Lee, K. (2012). Asexual and sexual developments of* Neurospora crassa *on natural substrata.* Fungal Ecology, *5(2), 223–229. doi:10.1016/j.funeco.2011.09.001.*

ACKNOWLEDGMENTS

Authors thank Drs. Martha Merrow, Daniel Shain, and Charot Rodeget for critical reading and insightful discussions. This work was supported by NIGMS-R15GM117523-01.

REFERENCES

Adamsson, M., Laike, T., & Morita, T. (2016). Annual variation in daily light exposure and circadian change of melatonin and cortisol concentrations at a northern latitude with large seasonal differences in photoperiod length. *Journal of Physiological Anthropology*, *36*(1), 6. https://doi.org/10.1186/s40101-016-0103-9.

Allebrandt, K. V., & Roenneberg, T. (2008). The search for circadian clock components in humans: New perspectives for association studies. *Brazilian Journal of Medical and Biological Research*, *41*(8), 716–721. https://doi.org/10.1590/S0100-879X2008000800013.

Alonso-Blanco, C., & Koornneef, M. (2000). Naturally occurring variation in Arabidopsis: An underexploited resource for plant genetics. *Trends in Plant Science*, *5*(1), 22–29.

Al-Saqur, A., & Smith, B. R. (1980). Surfactant resistance in natural Neurospora populations. *Neurospora Newsletter*, *27*, 17.

Aramayo, R., & Selker, E. U. (2013). Neurospora crassa, a model system for epigenetics research. *Cold Spring Harbor Perspectives in Biology*, *5*(10). https://doi.org/10.1101/cshperspect.a017921.

Aronson, B. D., Johnson, K. A., Loros, J. J., & Dunlap, J. C. (1994). Negative feedback defining a circadian clock: Autoregulation of the clock gene *frequency*. *Science*, *263*, 1578–1584.

Austin, B. (1968). An endogenous rhythm of spore discharge in Sordaria fimicola. *Annals of Botany*, *32*(2), 262–278. https://doi.org/10.1093/oxfordjournals.aob.a084207.

Baker, C. L., Loros, J. J., & Dunlap, J. C. (2012). The circadian clock of Neurospora crassa. *FEMS Microbiology Reviews*, *36*(1), 95–110. https://doi.org/10.1111/j.1574-6976.2011.00288.x.

Ballario, P., Vittorioso, P., Magrelli, A., Talora, C., Cabibbo, A., & Macino, G. (1996). White collar-1, a central regulator of blue light responses in Neurospora, is a zinc finger protein. *The EMBO Journal*, *15*(7), 1650–1657.

Beauchamp, P. M., Horn, E. W., & Gross, S. R. (1977). Proposed involvement of an internal promoter in regulation and synthesis of mitochondrial and cytoplasmic leucyl-tRNA synthetases of Neurospora. *Proceedings of the National Academy of Sciences of the United States of America*, *74*(3), 1172–1176.

Beaver, L. M., Gvakharia, B. O., Vollintine, T. S., Hege, D. M., Stanewsky, R., & Giebultowicz, J. M. (2002). Loss of circadian clock function decreases reproductive fitness in males of Drosophila melanogaster. *Proceedings of the National Academy of Sciences of the United States of America*, *99*(4), 2134–2139. https://doi.org/10.1073/pnas.032426699.

Belden, W. J., Larrondo, L. F., Froehlich, A. C., Shi, M., Chen, C. H., Loros, J. J., et al. (2007). The band mutation in Neurospora crassa is a dominant allele of ras-1 implicating RAS signaling in circadian output. *Genes & Development*, *21*(12), 1494–1505. https://doi.org/10.1101/gad.1551707.

Bell-Pedersen, D., Crosthwaite, S. K., Lakin-Thomas, P. L., Merrow, M., & Okland, M. (2001). The Neurospora circadian clock: Simple or complex? *Philosophical Transactions of the Royal Society of London. Series B, Biological Sciences*, *356*(1415), 1697–1709. https://doi.org/10.1098/rstb.2001.0968.

Bertrand, H., Chan, B. S., & Griffiths, A. J. (1985). Insertion of a foreign nucleotide sequence into mitochondrial DNA causes senescence in Neurospora intermedia. *Cell*, *41*(3), 877–884.

Bertrand, H., Griffiths, A. J., Court, D. A., & Cheng, C. K. (1986). An extrachromosomal plasmid is the etiological precursor of kalDNA insertion sequences in the mitochondrial chromosome of senescent Neurospora. *Cell*, *47*(5), 829–837.

Blakely, R. M., & Srb, A. M. (1962). Studies of the genetics and physiology of a nitrate non-utilizing strain of Neurospora. *Neurospora Newsletter, 2*(1), 5–6.
Bluhm, B. H., Burnham, A. M., & Dunkle, L. D. (2010). A circadian rhythm regulating hyphal melanization in Cercospora kikuchii. *Mycologia, 102*(6), 1221–1228. https://doi.org/10.3852/09-041.
Brandt, W. H. (1953). Zonation in a prolineless strain of Neurospora. *Mycologia, 45*(2), 194–208.
Brunner, M., & Schafmeier, T. (2006). Transcriptional and post-transcriptional regulation of the circadian clock of cyanobacteria and Neurospora. *Genes & Development, 20*(9), 1061–1074.
Campbell, J. L., & Turner, B. C. (1987). Recombination block in the Spore killer region of Neurospora. *Genome, 29*(1), 129–135.
Catcheside, D. G. (1975). Occurrence in wild strains of Neurospora crassa of genes controlling genetic recombination. *Australian Journal of Biological Sciences, 28*(2), 213–225.
Catcheside, D. E. (1981). Genes in Neurospora that suppress recombination when they are heterozygous. *Genetics, 98*(1), 55–76.
Chang, B., & Nakashima, H. (1997). Effects of light-dark cycles on the circadian conidiation rhythm in Neurospora crassa. *Journal of Plant Research, 110*(4), 449–453.
Chen, C. H., DeMay, B. S., Gladfelter, A. S., Dunlap, J. C., & Loros, J. J. (2010). Physical interaction between VIVID and white collar complex regulates photoadaptation in Neurospora. *Proceedings of the National Academy of Sciences of the United States of America, 107*(38), 16715–16720. https://doi.org/10.1073/pnas.1011190107.
Cheng, P., He, Q., He, Q., Wang, L., & Liu, Y. (2005). Regulation of the Neurospora circadian clock by an RNA helicase. *Genes & Development, 19*(2), 234–241. https://doi.org/10.1101/gad.1266805.
Cheng, P., He, Q., Yang, Y., Wang, L., & Liu, Y. (2003). Functional conservation of light, oxygen, or voltage domains in light sensing. *Proceedings of the National Academy of Sciences of the United States of America, 100*(10), 5938–5943.
Cheng, P., Yang, Y., Gardner, K. H., & Liu, Y. (2002). PAS domain-mediated WC-1/WC-2 interaction is essential for maintaining the steady-state level of WC-1 and the function of both proteins in circadian clock and light responses of Neurospora. *Molecular and Cellular Biology, 22*(2), 517–524.
Cheng, P., Yang, Y., & Liu, Y. (2001). Interlocked feedback loops contribute to the robustness of the Neurospora circadian clock. *Proceedings of the National Academy of Sciences of the United States of America, 98*(13), 7408–7413. https://doi.org/10.1073/pnas.121170298. 98/13/7408 [pii].
Cheng, P., Yang, Y., Wang, L., He, Q., & Liu, Y. (2003). WHITE COLLAR-1, a multifunctional neurospora protein involved in the circadian feedback loops, light sensing, and transcription repression of wc-2. *The Journal of Biological Chemistry, 278*(6), 3801–3808.
Christensen, M. K., Falkeid, G., Loros, J. J., Dunlap, J. C., Lillo, C., & Ruoff, P. (2004). A nitrate-induced frq-less oscillator in Neurospora crassa. *Journal of Biological Rhythms, 19*(4), 280–286. https://doi.org/10.1177/0748730404265532.
Collins, R. A., & Lambowitz, A. M. (1983). Structural variations and optional introns in the mitochondrial DNAs of Neurospora strains isolated from nature. *Plasmid, 9*(1), 53–70.
Colot, H. V., Park, G., Turner, G. E., Ringelberg, C., Crew, C. M., Litvinkova, L., et al. (2006). A high-throughput gene knockout procedure for Neurospora reveals functions for multiple transcription factors. *Proceedings of the National Academy of Sciences of the United States of America, 103*(27), 10352–10357. https://doi.org/10.1073/pnas.0601456103.
Coluccio, M. P., Sanchez, S. E., Kasulin, L., Yanovsky, M. J., & Botto, J. F. (2011). Genetic mapping of natural variation in a shade avoidance response: ELF3 is the candidate gene for a QTL in hypocotyl growth regulation. *Journal of Experimental Botany, 62*(1), 167–176. https://doi.org/10.1093/jxb/erq253.

Conrad, K. S., Hurley, J. M., Widom, J., Ringelberg, C. S., Loros, J. J., Dunlap, J. C., et al. (2016). Structure of the frequency-interacting RNA helicase: A protein interaction hub for the circadian clock. *The EMBO Journal*, *35*(15), 1707–1719. https://doi.org/10.15252/embj.201694327.

Correa, A., Lewis, Z. A., Greene, A. V., March, I. J., Gomer, R. H., & Bell-Pedersen, D. (2003). Multiple oscillators regulate circadian gene expression in Neurospora. *Proceedings of the National Academy of Sciences of the United States of America*, *100*(23), 13597–13602. https://doi.org/10.1073/pnas.2233734100.

Dawson, A., King, V. M., Bentley, G. E., & Ball, G. F. (2001). Photoperiodic control of seasonality in birds. *Journal of Biological Rhythms*, *16*(4), 365–380. https://doi.org/10.1177/074873001129002079.

Dettman, J. R., Anderson, J. B., & Kohn, L. M. (2008). Divergent adaptation promotes reproductive isolation among experimental populations of the filamentous fungus Neurospora. *BMC Evolutionary Biology*, *8*(1), 35. https://doi.org/10.1186/1471-2148-8-35.

Dettman, J. R., Anderson, J. B., & Kohn, L. M. (2010). Genome-wide investigation of reproductive isolation in experimental lineages and natural species of Neurospora: Identifying candidate regions by microarray-based genotyping and mapping. *Evolution*, *64*(3), 694–709. https://doi.org/10.1111/j.1558-5646.2009.00863.x.

Dettman, J. R., Harbinski, F. M., & Taylor, J. W. (2001). Ascospore morphology is a poor predictor of the phylogenetic relationships of Neurospora and Gelasinospora. *Fungal Genetics and Biology*, *34*(1), 49–61.

Dettman, J. R., Jacobson, D. J., & Taylor, J. W. (2003). A multilocus genealogical approach to phylogenetic species recognition in the model eukaryote Neurospora. *Evolution*, *57*(12), 2703–2720.

Dettman, J. R., Jacobson, D. J., & Taylor, J. W. (2006). Multilocus sequence data reveal extensive phylogenetic species diversity within the Neurospora discreta complex. *Mycologia*, *98*(3), 436–446.

Dettman, J. R., Jacobson, D. J., Turner, E., Pringle, A., & Taylor, J. W. (2003). Reproductive isolation and phylogenetic divergence in Neurospora: Comparing methods of species recognition in a model eukaryote. *Evolution*, *57*(12), 2721–2741.

Dodge, B. O. (1939). A new dominant lethal in Neurospora. *Journal of Heredity*, *30*, 467–474.

Dunlap, J. C., & Loros, J. J. (2017). Making time: Conservation of biological clocks from fungi to animals. *Microbiology spectrum*, *5*(3). FUNK-0039-2016, doi:10.1128/microbiolspec.FUNK-0039-2016.

Eberhart, B. M. (1961). Exogenous enzymes of Neurospora conidia and mycelia. *Journal of Cellular and Comparative Physiology*, *58*, 11–16.

Eberhart, B. M., Beck, R. S., & Goolsby, K. M. (1977). Cellulase of Neurospora crassa. *Journal of Bacteriology*, *130*(1), 181–186.

Edwards, K. D., Lynn, J. R., Gyula, P., Nagy, F., & Millar, A. J. (2005). Natural allelic variation in the temperature-compensation mechanisms of the Arabidopsis thaliana circadian clock. *Genetics*, *170*(1), 387–400. https://doi.org/10.1534/genetics.104.035238.

Ellison, C. E., Hall, C., Kowbel, D., Welch, J., Brem, R. B., Glass, N. L., et al. (2011). Population genomics and local adaptation in wild isolates of a model microbial eukaryote. *Proceedings of the National Academy of Sciences of the United States of America*, *108*(7), 2831–2836. https://doi.org/10.1073/pnas.1014971108.

Elvin, M., Loros, J. J., Dunlap, J. C., & Heintzen, C. (2005). The PAS/LOV protein VIVID supports a rapidly dampened daytime oscillator that facilitates entrainment of the Neurospora circadian clock. *Genes & Development*, *19*(21), 2593–2605.

Emerson, M. R. (1948). Chemical activation of ascospore germination in Neurospora crassa. *Journal of Bacteriology*, *55*(3), 327–330.

Forni, D., Pozzoli, U., Cagliani, R., Tresoldi, C., Menozzi, G., Riva, S., et al. (2014). Genetic adaptation of the human circadian clock to day-length latitudinal variations and relevance for affective disorders. *Genome Biology*, *15*(10), 499. https://doi.org/10.1186/s13059-014-0499-7.

Frederick, L., Uecker, F. A., & Benjamin, C. R. (1969). A new species of Neurospora from soil of West Pakistan. *Mycologia*, *61*(6), 1077–1084.

Friborg, O., Bjorvatn, B., Amponsah, B., & Pallesen, S. (2012). Associations between seasonal variations in day length (photoperiod), sleep timing, sleep quality and mood: A comparison between Ghana (5 degrees) and Norway (69 degrees). *Journal of Sleep Research*, *21*(2), 176–184. https://doi.org/10.1111/j.1365-2869.2011.00982.x.

Froehlich, A. C., Liu, Y., Loros, J. J., & Dunlap, J. C. (2002). White Collar-1, a circadian blue light photoreceptor, binding to the *frequency* promoter. *Science*, *297*(5582), 815–819.

Froehlich, A. C., Noh, B., Vierstra, R. D., Loros, J., & Dunlap, J. C. (2005). Genetic and molecular analysis of phytochromes from the filamentous fungus Neurospora crassa. *Eukaryotic Cell*, *4*(12), 2140–2152.

Garceau, N. Y., Liu, Y., Loros, J. J., & Dunlap, J. C. (1997). Alternative initiation of translation and time-specific phosphorylation yield multiple forms of the essential clock protein FREQUENCY. *Cell*, *89*(3), 469–476.

Gaston, K. J. (2000). Global patterns in biodiversity. *Nature*, *405*(6783), 220–227. https://doi.org/10.1038/35012228.

Gill, S., & Panda, S. (2011). Feeding mistiming decreases reproductive fitness in flies. *Cell Metabolism*, *13*(6), 613–614. https://doi.org/10.1016/j.cmet.2011.05.003.

Glossop, N. R. J., Lyons, L. C., & Hardin, P. E. (1999). Interlocked feedback loops within the *Drosophila* circadian oscillator. *Science*, *286*(5440), 766–768.

Goddard, D. R. (1935). The reversible heat activation inducing germination and increased respiration in the ascospores of Neurospora tetrasperma. *The Journal of General Physiology*, *19*(1), 45–60.

Goodrich-Tanrikulu, M., Stafford, A. E., & Jacobson, D. J. (1999). A systematic survey of the fatty acid composition of Neurospora strains. *Fungal Genetics Reports*, *46*(1) 3.

Görl, M., Merrow, M., Huttner, B., Johnson, J., Roenneberg, T., & Brunner, M. (2001). A PEST-like element in FREQUENCY determines the length of the circadian period in Neurospora crassa. *The EMBO Journal*, *20*(24), 7074–7084.

Granshaw, T., Tsukamoto, M., & Brody, S. (2003). Circadian rhythms in Neurospora crassa: Farnesol or geraniol allow expression of rhythmicity in the otherwise arrhythmic strains frq10, wc-1, and wc-2. *Journal of Biological Rhythms*, *18*(4), 287–296. https://doi.org/10.1177/0748730403255934.

Green, R. M., Tingay, S., Wang, Z. Y., & Tobin, E. M. (2002). Circadian rhythms confer a higher level of fitness to Arabidopsis plants. *Plant Physiology*, *129*(2), 576–584. https://doi.org/10.1104/pp.004374.

Greene, A. V., Keller, N., Haas, H., & Bell-Pedersen, D. (2003). A circadian oscillator in Aspergillus spp. regulates daily development and gene expression. *Eukaryotic Cell*, *2*(2), 231–237.

Griffiths, A. J. (1992). Fungal senescence. *Annual Review of Genetics*, *26*(1), 351–372. https://doi.org/10.1146/annurev.ge.26.120192.002031.

Griffiths, A. J. (1995). Natural plasmids of filamentous fungi. *Microbiological Reviews*, *59*(4), 673–685.

Griffiths, A. J., & Bertrand, H. (1984). Unstable cytoplasms in Hawaiian strains of Neurospora intermedia. *Current Genetics*, *8*(5), 387–398. https://doi.org/10.1007/BF00419828.

Grindle, M., & Davis, R. H. (1970). Variations in enzyme phenotypes and their underlying genotypes among wild strains of the genus Neurospora. *Genetical Research*, *15*(1), 99–107.

Grove, G., & Marzluf, G. A. (1981). Identification of the product of the major regulatory gene of the nitrogen control circuit of Neurospora crassa as a nuclear DNA-binding protein. *The Journal of Biological Chemistry*, *256*(1), 463–470.

Guo, J., Cheng, P., & Liu, Y. (2010). Functional significance of FRH in regulating the phosphorylation and stability of Neurospora circadian clock protein FRQ. *The Journal of Biological Chemistry*, *285*(15), 11508–11515. https://doi.org/10.1074/jbc.M109.071688.
Hanson, M. A., & Marzluf, G. A. (1975). Control of the synthesis of a single enzyme by multiple regulatory circuits in Neurospora crassa. *Proceedings of the National Academy of Sciences of the United States of America*, *72*(4), 1240–1244.
Hau, M. (2001). Timing of breeding in variable environments: Tropical birds as model systems. *Hormones and Behavior*, *40*(2), 281–290. https://doi.org/10.1006/hbeh.2001.1673.
He, Q., Cha, J., He, Q., Lee, H. C., Yang, Y., & Liu, Y. (2006). CKI and CKII mediate the FREQUENCY-dependent phosphorylation of the WHITE COLLAR complex to close the Neurospora circadian negative feedback loop. *Genes & Development*, *20*(18), 2552–2565. https://doi.org/10.1101/gad.1463506.
He, Q., Cheng, P., Yang, Y., Wang, L., Gardner, K. H., & Liu, Y. (2002). White Collar-1, a DNA binding transcription factor and a light sensor. *Science*, *297*(5582), 840–843.
He, Q., Cheng, P., Yang, Y., Yu, H., & Liu, Y. (2003). FWD1-mediated degradation of FREQUENCY in Neurospora establishes a conserved mechanism for circadian clock regulation. *The EMBO Journal*, *22*(17), 4421–4430.
He, Q., & Liu, Y. (2005a). Degradation of the Neurospora circadian clock protein FREQUENCY through the ubiquitin-proteasome pathway. *Biochemical Society Transactions*, *33*(Pt 5), 953–956.
He, Q., & Liu, Y. (2005b). Molecular mechanism of light responses in Neurospora: From light-induced transcription to photoadaptation. *Genes & Development*, *19*(23), 2888–2899.
He, C., Nastasja de, G., Bok, J. W., & Griffiths, A. J. (2000). Kalilo plasmids are a family of four distinct members with individual global distributions across species. *Current Genetics*, *37*(1), 39–44.
Heintzen, C., & Liu, Y. (2007). The Neurospora crassa circadian clock. *Advances in Genetics*, *58*, 25–66. https://doi.org/10.1016/S0065-2660(06)58002-2.
Heintzen, C., Loros, J. J., & Dunlap, J. C. (2001). The PAS protein VIVID defines a clock-associated feedback loop that represses light input, modulates gating, and regulates clock resetting. *Cell*, *104*(3), 453–464.
Hevia, M. A., Canessa, P., & Larrondo, L. F. (2016). Circadian clocks and the regulation of virulence in fungi: Getting up to speed. *Seminars in Cell & Developmental Biology*, *57*, 147–155. https://doi.org/10.1016/j.semcdb.2016.03.021.
Hevia, M. A., Canessa, P., Muller-Esparza, H., & Larrondo, L. F. (2015). A circadian oscillator in the fungus Botrytis cinerea regulates virulence when infecting Arabidopsis thaliana. *Proceedings of the National Academy of Sciences of the United States of America*, *112*(28), 8744–8749. https://doi.org/10.1073/pnas.1508432112.
Horowitz, N. H., Fling, M., Macleod, H., & Sueoka, N. (1961). A genetic study of two new structural forms of tyrosinase in Neurospora. *Genetics*, *46*(8), 1015–1024.
Hunt, S. M., Thompson, S., Elvin, M., & Heintzen, C. (2010). VIVID interacts with the WHITE COLLAR complex and FREQUENCY-interacting RNA helicase to alter light and clock responses in Neurospora. *Proceedings of the National Academy of Sciences of the United States of America*, *107*(38), 16709–16714. https://doi.org/10.1073/pnas.1009474107.
Hurley, J. M., Larrondo, L. F., Loros, J. J., & Dunlap, J. C. (2013). Conserved RNA helicase FRH acts nonenzymatically to support the intrinsically disordered neurospora clock protein FRQ. *Molecular Cell*, *52*(6), 832–843. https://doi.org/10.1016/j.molcel.2013.11.005.
Hurley, J., Loros, J. J., & Dunlap, J. C. (2015). Dissecting the mechanisms of the clock in Neurospora. *Methods in Enzymology*, *551*, 29–52. https://doi.org/10.1016/bs.mie.2014.10.009.
Jacobson, D. J., Dettman, J. R., Adams, R. I., Boesl, C., Sultana, S., Roenneberg, T., et al. (2006). New findings of Neurospora in Europe and comparisons of diversity in temperate climates on continental scales. *Mycologia*, *98*(4), 550–559.

Jacobson, D. J., Powell, A. J., Dettman, J. R., Saenz, G. S., Barton, M. M., Hiltz, M. D., et al. (2004). Neurospora in temperate forests of western North America. *Mycologia, 96*(1), 66–74.
Kaldi, K., Gonzalez, B. H., & Brunner, M. (2006). Transcriptional regulation of the Neurospora circadian clock gene wc-1 affects the phase of circadian output. *EMBO Reports*, 7(2), 199–204.
Kim, T. S., Booth, J. G., Gauch, H. G., Jr., Sun, Q., Park, J., Lee, Y. H., et al. (2008). Simple sequence repeats in Neurospora crassa: Distribution, polymorphism and evolutionary inference. *BMC Genomics, 9*(1), 31. https://doi.org/10.1186/1471-2164-9-31.
Kim, T. S., Logsdon, B. A., Park, S., Mezey, J. G., & Lee, K. (2007). Quantitative trait loci for the circadian clock in Neurospora crassa. *Genetics, 177*(4), 2335–2347. https://doi.org/10.1534/genetics.107.077958.
Lakin-Thomas, P. L. (1998). Choline depletion, frq mutations, and temperature compensation of the circadian rhythm in Neurospora crassa. *Journal of Biological Rhythms, 13*(4), 268–277. https://doi.org/10.1177/074873098129000101.
Lakin-Thomas, P. L., Bell-Pedersen, D., & Brody, S. (2011). The genetics of circadian rhythms in Neurospora. *Advances in Genetics, 74*, 55–103. https://doi.org/10.1016/B978-0-12-387690-4.00003-9.
Lakin-Thomas, P. L., & Brody, S. (2000). Circadian rhythms in Neurospora crassa: Lipid deficiencies restore robust rhythmicity to null frequency and white-collar mutants. *Proceedings of the National Academy of Sciences of the United States of America, 97*(1), 256–261.
Lankinen, P., & Forsman, P. (2006). Independence of genetic geographical variation between photoperiodic diapause, circadian eclosion rhythm, and Thr-Gly repeat region of the period gene in Drosophila littoralis. *Journal of Biological Rhythms, 21*(1), 3–12. https://doi.org/10.1177/0748730405283418.
Lauinger, L., Diernfellner, A., Falk, S., & Brunner, M. (2014). The RNA helicase FRH is an ATP-dependent regulator of CK1a in the circadian clock of Neurospora crassa. *Nature Communications, 5*, 3598. https://doi.org/10.1038/ncomms4598.
Lee, K. (2012). Asexual and sexual developments of Neurospora crassa on natural substrata. *Fungal Ecology, 5*(2), 223–229. https://doi.org/10.1016/j.funeco.2011.09.001.
Lee, K., & Dighton, J. (2013). Advancement of functional genomics of a model species of Neurospora and its use for ecological genomics of soil fungi. In B. A. Horwitz, P. K. Mukherjee, M. Mukherjee, & C. P. Kubicek (Eds.), *Genomics of soil- and plant-associated fungi: Vol. 36.* (pp. 29–44). Berlin, Heidelberg: Springer.
Lee, K., Dunlap, C. J., & Loros, J. J. (2003). Roles for WHITE COLLAR-1 in circadian and general photoperception in *Neurospora crassa*. *Genetics, 163*(1), 103–114.
Lee, K., Loros, J. J., & Dunlap, J. C. (2000). Interconnected feedback loops in the *Neurospora* circadian system. *Science, 289*(5476), 107–110.
Lempe, J., Balasubramanian, S., Sureshkumar, S., Singh, A., Schmid, M., & Weigel, D. (2005). Diversity of flowering responses in wild Arabidopsis thaliana strains. *PLoS Genetics, 1*(1), 109–118. https://doi.org/10.1371/journal.pgen.0010006.
Leslie, J. F., & Raju, N. B. (1985). Recessive mutations from natural populations of Neurospora crassa that are expressed in the sexual diplophase. *Genetics, 111*(4), 759–777.
Levine, W. B., & Marzluf, G. A. (1989). Isolation and characterization of cadmium-resistant mutants of Neurospora crassa. *Canadian Journal of Microbiology, 35*(3), 359–365.
Lewis, M. T., & Feldman, J. F. (1996). Evolution of the frequency (frq) clock locus in Ascomycete fungi. *Molecular Biology and Evolution, 13*(9), 1233–1241.
Li, N., Joska, T. M., Ruesch, C. E., Coster, S. J., & Belden, W. J. (2015). The frequency natural antisense transcript first promotes, then represses, frequency gene expression via facultative heterochromatin. *Proceedings of the National Academy of Sciences of the United States of America, 112*(14), 4357–4362. https://doi.org/10.1073/pnas.1406130112.

Liu, Y., Loros, J., & Dunlap, J. C. (2000). Phosphorylation of the Neurospora clock protein FREQUENCY determines its degradation rate and strongly influences the period length of the circadian clock. *Proceedings of the National Academy of Sciences of the United States of America*, *97*(1), 234–239.

Loros, J. J., Richman, A., & Feldman, J. F. (1986). A recessive circadian clock mutation at the frq locus of Neurospora crassa. *Genetics*, *114*(4), 1095–1110.

Low, K. H., Chen, W. F., Yildirim, E., & Edery, I. (2012). Natural variation in the Drosophila melanogaster clock gene period modulates splicing of its 3′-terminal intron and mid-day siesta. *PLoS One*, 7(11), e49536. https://doi.org/10.1371/journal.pone.0049536.

Low, K. H., Lim, C., Ko, H. W., & Edery, I. (2008). Natural variation in the splice site strength of a clock gene and species-specific thermal adaptation. *Neuron*, *60*(6), 1054–1067. https://doi.org/10.1016/j.neuron.2008.10.048.

Mahadevan, P. R., & Eberhart, B. (1964). Aryl beta-glucosidase of some Neurospora strains. *Biochimica et Biophysica Acta*, *90*, 214–215.

Maheshwari, R., & Navaraj, A. (2008). Senescence in fungi: The view from Neurospora. *FEMS Microbiology Letters*, *280*(2), 135–143. https://doi.org/10.1111/j.1574-6968.2007.01027.x.

Mannella, C. A., Pittenger, T. H., & Lambowitz, A. M. (1979). Transmission of mitochondrial deoxyribonucleic acid in Neurospora crassa sexual crosses. *Journal of Bacteriology*, *137*(3), 1449–1451.

McCluskey, K., Wiest, A., & Plamann, M. (2010). The Fungal Genetics Stock Center: A repository for 50 years of fungal genetics research. *Journal of Biosciences*, *35*(1), 119–126.

Mehra, A., Shi, M., Baker, C. L., Colot, H. V., Loros, J. J., & Dunlap, J. C. (2009). A role for casein kinase 2 in the mechanism underlying circadian temperature compensation. *Cell*, *137*(4), 749–760. https://doi.org/10.1016/j.cell.2009.03.019.

Merrow, M., Brunner, M., & Roenneberg, T. (1999). Assignment of circadian function for the Neurospora clock gene frequency. *Nature*, *399*(6736), 584–586. https://doi.org/10.1038/21190.

Merrow, M. W., & Dunlap, J. C. (1994). Intergeneric complementation of a circadian rhythmicity defect: Phylogenetic conservation of structure and function of the clock gene frequency. *The EMBO Journal*, *13*(10), 2257–2266.

Merrow, M. W., Garceau, N. Y., & Dunlap, J. C. (1997). Dissection of a circadian oscillation into discrete domains. *Proceedings of the National Academy of Sciences of the United States of America*, *94*(8), 3877–3882.

Metzenberg, R. L., & Ahlgren, S. K. (1971). Structural and regulatory control of aryl sulfatase in Neurospora: The use of interspecific differences in structural genes. *Genetics*, *68*(3), 369–381.

Michael, T. P., Park, S., Kim, T. S., Booth, J., Byer, A., Sun, Q., et al. (2007). Simple sequence repeats provide a substrate for phenotypic variation in the Neurospora crassa circadian clock. *PLoS One*, *2*(8), e795. https://doi.org/10.1371/journal.pone.0000795.

Michael, T. P., Salome, P. A., Yu, H. J., Spencer, T. R., Sharp, E. L., McPeek, M. A., et al. (2003). Enhanced fitness conferred by naturally occurring variation in the circadian clock. *Science*, *302*(5647), 1049–1053. https://doi.org/10.1126/science.1082971.

Miller, B. H., Olson, S. L., Turek, F. W., Levine, J. E., Horton, T. H., & Takahashi, J. S. (2004). Circadian clock mutation disrupts estrous cyclicity and maintenance of pregnancy. *Current Biology*, *14*(15), 1367–1373. https://doi.org/10.1016/j.cub.2004.07.055.

Morgan, L. W., & Feldman, J. F. (1998). Preliminary investigation of the circadian rhythms of wild-collected Neurospora strains. *Fungal Genetics Newsletter*, *45*, 30–31.

Mylyk, O. M. (1975). Heterokaryon incompatibility genes in Neurospora crassa detected using duplication-producing chromosome rearrangements. *Genetics*, *80*(1), 107–124.

Mylyk, O. M. (1976). Heteromorphism for heterokaryon incompatibility genes in natural populations of Neurospora Crassa. *Genetics*, *83*(2), 275–284.
Natvig, D. O., May, G., & Taylor, J. W. (1984). Distribution and evolutionary significance of mitochondrial plasmids in Neurospora spp. *Journal of Bacteriology*, *159*(1), 288–293.
Navaraj, A., Pandit, A., & Maheshwari, R. (2000). Senescent: A new Neurospora crassa nuclear gene mutant derived from nature exhibits mitochondrial abnormalities and a "death" phenotype. *Fungal Genetics and Biology*, *29*(3), 165–173. https://doi.org/10.1006/fgbi.2000.1193.
Newmeyer, D. (1970). A suppressor of the heterokaryon-incompatibility associated with mating type in Neurospora crassa. *Canadian Journal of Genetics and Cytology*, *12*(4), 914–926.
Nsa, I. Y., Karunarathna, N., Liu, X., Huang, H., Boetteger, B., & Bell-Pedersen, D. (2015). A novel cryptochrome-dependent oscillator in Neurospora crassa. *Genetics*, *199*(1), 233–245. https://doi.org/10.1534/genetics.114.169441.
Oliveira, A. G., Stevani, C. V., Waldenmaier, H. E., Viviani, V., Emerson, J. M., Loros, J. J., et al. (2015). Circadian control sheds light on fungal bioluminescence. *Current Biology*, *25*(7), 964–968. https://doi.org/10.1016/j.cub.2015.02.021.
Park, S., & Lee, K. (2004). Inverted race tube assay for a circadian clock studies of the Neurospora accessions. *Fungal Genetics Newsletter*, *51*(5), 12–14.
Perkins, D. D. (1975). The use of duplication-generating rearrangements for studying heterokaryon incompatibility genes in Neurospora. *Genetics*, *80*(1), 87–105.
Perkins, D. D., Metzenberg, R. L., Raju, N. B., Selker, E. U., & Barry, E. G. (1986). Reversal of a Neurospora translocation by crossing over involving displaced rDNA, and methylation of the rDNA segments that result from recombination. *Genetics*, *114*(3), 791–817.
Perkins, D. D., Radford, A., Newmeyer, D., & Bjorkman, M. (1982). Chromosomal loci of Neurospora crassa. *Microbiological Reviews*, *46*(4), 426–570.
Perkins, D. D., & Turner, B. C. (1988). Neurospora from natural populations: Toward the population biology of a haploid eukaryote. *Experimental Mycology*, *12*(2), 91–131.
Perkins, D. D., Turner, B. C., & Barry, E. G. (1976). Strains of Neurospora collected from nature. *Evolution*, *30*(2), 281–313.
Pittendrigh, C. S., Bruce, V. G., Rosensweig, N. S., & Rubin, M. L. (1959). Growth patterns in Neurospora: A biological clock in Neurospora. *Nature*, *184*(4681), 169–170. https://doi.org/10.1038/184169a0.
Pittendrigh, C. S., Kyner, W. T., & Takamura, T. (1991). The amplitude of circadian oscillations: Temperature dependence, latitudinal clines, and the photoperiodic time measurement. *Journal of Biological Rhythms*, *6*(4), 299–313. https://doi.org/10.1177/074873049100600402.
Pittendrigh, C. S., & Takamura, T. (1989). Latitudinal clines in the properties of a circadian pacemaker. *Journal of Biological Rhythms*, *4*(2), 217–235.
Pregueiro, A. M., Liu, Q., Baker, C. L., Dunlap, J. C., & Loros, J. J. (2006). The Neurospora checkpoint kinase 2: A regulatory link between the circadian and cell cycles. *Science*, *313*(5787), 644–649.
Press, M. O., Lanctot, A., & Queitsch, C. (2016). PIF4 and ELF3 act independently in Arabidopsis thaliana thermoresponsive flowering. *PLoS One*, *11*(8), e0161791. https://doi.org/10.1371/journal.pone.0161791.
Pringle, A., & Taylor, J. (2002). The fitness of filamentous fungi. *Trends in Microbiology*, *10*(10), 474–481.
Querfurth, C., Diernfellner, A. C., Gin, E., Malzahn, E., Hofer, T., & Brunner, M. (2011). Circadian conformational change of the Neurospora clock protein FREQUENCY triggered by clustered hyperphosphorylation of a basic domain. *Molecular Cell*, *43*(5), 713–722. https://doi.org/10.1016/j.molcel.2011.06.033.

Raju, N. B., & Leslie, J. F. (1992). Cytology of recessive sexual-phase mutants from wild strains of Neurospora crassa. *Genome*, *35*(5), 815–826.
Raschke, A., Ibanez, C., Ullrich, K. K., Anwer, M. U., Becker, S., Glockner, A., et al. (2015). Natural variants of ELF3 affect thermomorphogenesis by transcriptionally modulating PIF4-dependent auxin response genes. *BMC Plant Biology*, *15*(1), 197. https://doi.org/10.1186/s12870-015-0566-6.
Reddy, M. M., & Threlkeld, S. F. (1971a). Genetic studies of isozymes in Neurospora. I. A study of eight species. *Canadian Journal of Genetics and Cytology*, *13*(2), 298–305.
Reddy, M. M., & Threlkeld, S. F. (1971b). Isozyme variation in natural populations. *Fungal Genetics Reports*, *18*(1), 7.
Reddy, M. M., & Threlkeld, S. F. (1972). Genetic studies of isozymes in Neurospora. II. Esterases. *Canadian Journal of Genetics and Cytology*, *14*(3), 527–531.
Reierth, E., Van't Hof, T. J., & Stokkan, K. A. (1999). Seasonal and daily variations in plasma melatonin in the high-arctic Svalbard ptarmigan (Lagopus mutus hyperboreus). *Journal of Biological Rhythms*, *14*(4), 314–319. https://doi.org/10.1177/074873099129000731.
Rieck, A., Griffiths, A. J., & Bertrand, H. (1982). Mitochondrial variants of Neurospora intermedia from nature. *Canadian Journal of Genetics and Cytology*, *24*(6), 741–759.
Roenneberg, T., & Taylor, W. (2000). Automated recordings of bioluminescence with special reference to the analysis of circadian rhythms. *Methods in Enzymology*, *305*, 104–119.
Romero-Olivares, A. L., Taylor, J. W., & Treseder, K. K. (2015). Neurospora discreta as a model to assess adaptation of soil fungi to warming. *BMC Evolutionary Biology*, *15*(1), 198. https://doi.org/10.1186/s12862-015-0482-2.
Rosato, E., & Kyriacou, C. P. (2011). The role of natural selection in circadian behaviour: A molecular-genetic approach. *Essays in Biochemistry*, *49*(1), 71–85. https://doi.org/10.1042/BSE0490071.
Ruegg, C., Ammer, D., & Lerch, K. (1982). Comparison of amino-acid-sequence and thermostability of tyrosinase from 3 wild-type strains of Neurospora crassa. *Journal of Biological Chemistry*, *257*(11), 6420–6426.
Salichos, L., & Rokas, A. (2010). The diversity and evolution of circadian clock proteins in fungi. *Mycologia*, *102*(2), 269–278.
Sargent, M. L., & Kaltenborn, S. H. (1972). Effects of medium composition and carbon dioxide on circadian conidiation in neurospora. *Plant Physiology*, *50*(1), 171–175.
Sargent, M. L., & Woodward, D. O. (1969). Genetic determinants of circadian rhythmicity in Neurospora. *Journal of Bacteriology*, *97*(2), 861–866.
Sawyer, L. A., Hennessy, J. M., Peixoto, A. A., Rosato, E., Parkinson, H., Costa, R., et al. (1997). Natural variation in a Drosophila clock gene and temperature compensation. *Science*, *278*(5346), 2117–2120.
Schafmeier, T., Haase, A., Káldi, K., Scholz, J., Fuchs, M., & Brunner, M. (2005). Transcriptional feedback of neurospora circadian clock gene by phosphorylation-dependent inactivation of its transcription factor. *Cell*, *122*(2), 235–246.
Schneider, K., Perrino, S., Oelhafen, K., Li, S., Zatsepin, A., Lakin-Thomas, P., et al. (2009). Rhythmic conidiation in constant light in vivid mutants of Neurospora crassa. *Genetics*, *181*(3), 917–931. https://doi.org/10.1534/genetics.108.097808.
Schwerdtfeger, C., & Linden, H. (2003). VIVID is a flavoprotein and serves as a fungal blue light photoreceptor for photoadaptation. *The EMBO Journal*, *22*(18), 4846–4855.
Selker, E. U., & Stevens, J. N. (1987). Signal for DNA methylation associated with tandem duplication in Neurospora crassa. *Molecular and Cellular Biology*, 7(3), 1032–1038.
Shear, C. L., & Dodge, B. O. (1927). Life histories and heterothallism of the red bread-mold fungi of the Monilia sitophila group. *Journal of Agricultural Research*, *34*(11), 1019–1042.
Shearman, L. P., Sriram, S., Weaver, D. R., Maywood, E. S., Chaves, I. e., Zheng, B., et al. (2000). Interacting molecular loops in the mammalian circadian clock. *Science*, *288*(5468), 1013–1019.

Shi, M., Collett, M., Loros, J. J., & Dunlap, J. C. (2010). FRQ-interacting RNA helicase mediates negative and positive feedback in the Neurospora circadian clock. *Genetics*, *184*(2), 351–361. https://doi.org/10.1534/genetics.109.111393.

Shimomura, K., Low-Zeddies, S. S., King, D. P., Steeves, T. D., Whiteley, A., Kushla, J., et al. (2001). Genome-wide epistatic interaction analysis reveals complex genetic determinants of circadian behavior in mice. *Genome Research*, *11*(6), 959–980. https://doi.org/10.1101/gr.171601.

Shrode, L. B., Lewis, Z. A., White, L. D., Bell-Pedersen, D., & Ebbole, D. J. (2001). vvd is required for light adaptation of conidiation-specific genes of Neurospora crassa, but not circadian conidiation. *Fungal Genetics and Biology*, *32*(3), 169–181. https://doi.org/10.1006/fgbi.2001.1264. S1087-1845(01)91264-2 [pii].

Skupski, M. P., Jackson, D. A., & Natvig, D. O. (1997). Phylogenetic analysis of heterothallic Neurospora species. *Fungal Genetics and Biology*, *21*(1), 153–162.

Smith, D. A. (1975). A mutant affecting meiosis in Neurospora. *Genetics*, *80*(1), 125–133.

Smith, M. L., Bruhn, J. N., & Anderson, J. B. (1992). The fungus Armillaria bulbosa is among the largest and oldest living organisms. *Nature*, *356*(6368), 428.

Spieth, P. T. (1975). Population genetics of allozyme variation in Neurospora intermedia. *Genetics*, *80*(4), 785–805.

Springer, M. L. (1993). Genetic control of fungal differentiation: The three sporulation pathways of Neurospora crassa. *BioEssays*, *15*(6), 365–374.

Springer, M. L., & Yanofsky, C. (1989). A morphological and genetic analysis of conidiophore development in Neurospora crassa. *Genes & Development*, *3*(4), 559–571.

Svetec, N., Zhao, L., Saelao, P., Chiu, J. C., & Begun, D. J. (2015). Evidence that natural selection maintains genetic variation for sleep in Drosophila melanogaster. *BMC Evolutionary Biology*, *15*(1), 41. https://doi.org/10.1186/s12862-015-0316-2.

Swarup, K., Alonso-Blanco, C., Lynn, J. R., Michaels, S. D., Amasino, R. M., Koornneef, M., et al. (1999). Natural allelic variation identifies new genes in the Arabidopsis circadian system. *The Plant Journal*, *20*(1), 67–77.

Tai, F. L. (1935). Two new species of Neurospora. *Mycologia*, *27*(3), 328–330.

Talora, C., Franchi, L., Linden, H., Ballario, P., & Macino, G. (1999). Role of a white collar-1-white collar-2 complex in blue-light signal transduction. *The EMBO Journal*, *18*(18), 4961–4968. https://doi.org/10.1093/emboj/18.18.4961.

Taylor, J. W. (1986). Fungal evolutionary biology and mitochondrial-DNA. *Experimental Mycology*, *10*(4), 259–269. https://doi.org/10.1016/0147-5975(86)90011-3.

Taylor, J. W., Koupel, T., & Whitmer, S. (1987). *Distribution of mitochondrial DNA length mutations in Neurospora crassa isolates from Florida and the Ivory Coast (abstract)*. Paper presented at the Mycological Society of America Newsletter.

Taylor, J. W., Smolich, B. D., & May, G. (1985). An evolutionary comparison of homologous mitochondrial plasmid DNAs from three Neurospora species. *Molecular & General Genetics*, *201*(2), 161–167.

Taylor, J. W., Smolich, B. D., & May, G. (1986). Evolution and mitochondrial-DNA in Neurospora crassa. *Evolution*, *40*(4), 716–739. https://doi.org/10.2307/2408458.

Thom, C., & Ayers, S. H. (1916). Effect of pasteurization on mold spores. *Journal of Agricultural Research*, *6*(4), 153–166.

Toyota, K., Onai, K., & Nakashima, H. (2002). A new *wc-1* mutant of *Neurospora crassa* shows unique light sensitivity in the circadian conidiation rhythm. *Molecular Genetics and Genomics*, *268*(1), 56–61.

Traeger, S., & Nowrousian, M. (2015). Analysis of circadian rhythms in the basal filamentous ascomycete Pyronema confluens. *G3: Genes, Genomes, Genetics*, *5*(10), 2061–2071. https://doi.org/10.1534/g3.115.020461.

Turner, B. C., Perkins, D. D., & Fairfield, A. (2001). Neurospora from natural populations: A global study. *Fungal Genetics and Biology*, *32*(2), 67–92. https://doi.org/10.1006/fgbi.2001.1247.

Tuveson, R. W., & Peterson, J. F. (1972). Virus-like particles in certain slow-growing strains of Neurospora crassa. *Virology*, *47*(2), 527–531.
Wang, B., Kettenbach, A. N., Gerber, S. A., Loros, J. J., & Dunlap, J. C. (2014). Neurospora WC-1 recruits SWI/SNF to remodel frequency and initiate a circadian cycle. *PLoS Genetics*, *10*(9), e1004599. https://doi.org/10.1371/journal.pgen.1004599.
Weiss, B., & Turian, G. (1966). A study of conidiation in Neurospora crassa. *Journal of General Microbiology*, *44*(3), 407–418. https://doi.org/10.1099/00221287-44-3-407.
Westergaard, M., & Mitchell, H. K. (1947). Neurospora V. A synthetic medium favoring sexual reproduction. *American Journal of Botany*, *34*(10), 573–577.
Xu, K., DiAngelo, J. R., Hughes, M. E., Hogenesch, J. B., & Sehgal, A. (2011). The circadian clock interacts with metabolic physiology to influence reproductive fitness. *Cell Metabolism*, *13*(6), 639–654. https://doi.org/10.1016/j.cmet.2011.05.001.
Xue, Z., Ye, Q., Anson, S. R., Yang, J., Xiao, G., Kowbel, D., et al. (2014). Transcriptional interference by antisense RNA is required for circadian clock function. *Nature*, *514*(7524), 650–653. https://doi.org/10.1038/nature13671.
Yang, Y., Cheng, P., & Liu, Y. (2002). Regulation of the Neurospora circadian clock by casein kinase II. *Genes & Development*, *16*(8), 994–1006.
Yang, Y., Cheng, P., Zhi, G., & Liu, Y. (2001). Identification of a calcium/calmodulin-dependent protein kinase that phosphorylates the Neurospora circadian clock protein FREQUENCY. *The Journal of Biological Chemistry*, *276*(44), 41064–41072.
Yang, Y., He, Q., Cheng, P., Wrage, P., Yarden, O., & Liu, Y. (2004). Distinct roles for PP1 and PP2A in the Neurospora circadian clock. *Genes & Development*, *18*(3), 255–260.
Young, J. L., & Marzluf, G. A. (1991). Molecular comparison of the negative-acting nitrogen control gene, nmr, in Neurospora crassa and other Neurospora and fungal species. *Biochemical Genetics*, *29*(9–10), 447–459.
Yu, S. A., Garrett, K., & Sussman, A. S. (1971). Genetic control of multiple forms of trehalase in Neurospora crassa. *Genetics*, *68*(4), 473–481.
Zalokar, M. (1954). Studies on biosynthesis of carotenoids in Neurospora crassa. *Archives of Biochemistry and Biophysics*, *50*(1), 71–80.
Zalokar, M. (1959). Enzyme activity and cell differentiation in Neurospora. *American Journal of Botany*, *46*(7), 555–559.
Zalokar, M., & Cochrane, V. W. (1956). Diphosphopyridine nucleotidase in the life cycle of Neurospora crassa. *American Journal of Botany*, *43*(2), 107–110.
Zoltowski, B. D., & Crane, B. R. (2008). Light activation of the LOV protein vivid generates a rapidly exchanging dimer. *Biochemistry*, *47*(27), 7012–7019. https://doi.org/10.1021/bi8007017.

CHAPTER TWO

Natural Variation and Genetics of Photoperiodism in *Wyeomyia smithii*

William E. Bradshaw[1], Christina M. Holzapfel
Institute of Ecology and Evolution, University of Oregon, Eugene, OR, United States
[1]Corresponding author: e-mail address: mosquito@uoregon.edu

Contents

Abstract

Seasonal change in the temperate and polar regions of Earth determines how the world looks around us and, in fact, how we live our day-to-day lives. For biological organisms, seasonal change typically involves complex physiological and metabolic reorganization, the majority of which is regulated by photoperiodism. Photoperiodism is the ability of animals and plants to use day length or night length, resulting in life-historical transformations, including seasonal development, migration, reproduction, and dormancy. Seasonal timing determines not only survival and reproductive success but also the structure and organization of complex communities and, ultimately, the biomes of Earth. Herein, a small mosquito, *Wyeomyia smithii*, that lives only in the water-filled leaves of a carnivorous plant over a wide geographic range, is used to explore the

Advances in Genetics, Volume 99
ISSN 0065-2660
https://doi.org/10.1016/bs.adgen.2017.09.002

genetic and evolutionary basis of photoperiodism. Photoperiodism in *W. smithii* is considered in the context of its historical biogeography in nature to examine the startling finding that recent rapid climate change can drive genetic change in plants and animals at break-neck speed, and to challenge the ponderous 80+ year search for connections between daily and seasonal time-keeping mechanisms. Finally, a model is proposed that reconciles the seemingly disparate 24-h daily clock driven by the invariant rotation of Earth about its axis with the evolutionarily flexible seasonal timer orchestrated by variable seasonality driven by the rotation of Earth about the Sun.

1. INTRODUCTION

1.1 Raison d'être of Photoperiodism

Photoperiodism is essential for the maintenance of plant and animal fitness in temperate and arctic climates.

Photoperiodism is the ability of plants and animals to use the length of day or night, resulting in modification of their activities (Bradshaw & Holzapfel, 2007; Kubota et al., 2014; Lucas-Reina, Romero, Romero, & Valverde, 2015). Photoperiodism orchestrates such seasonal activities as growth, development, reproduction, migration, and dormancy that make a direct contribution to survivorship and reproductive success. Hence, being at the right place at the right time of year is essential for optimizing fitness at temperate latitudes. The trick is being able to anticipate and prepare in advance for seasonal change. "Food, temperature, moisture, and air pressure are all very much less regular in their seasonal procession and are therefore likely to be less effective as "clocks" than photoperiod" (Andrewartha & Birch, 1954, p. 294). Indeed, in comparison with temperature (Fig. 1) the length of day, or photoperiod, fluctuates regularly with the changing seasons and, at any given spot on Earth, is the same today as it was 10,000 years ago and will be 10,000 years into the future. Hence, photoperiod constitutes the most reliable environmental cue for predicting future seasonal change through evolutionary time. It should be no surprise that plants from algae to angiosperms and animals from rotifers to rodents use photoperiodism to regulate their seasonal activities (Nelson, Denlinger, & Somers, 2010).

1.2 Geographic Clines

Geographically variable climates impose concomitant selective pressure on the timing of seasonal activities and, hence, on photoperiodic response.

Length of the growing season favorable for growth, development, and reproduction decreases with increasing latitude and altitude; at the same

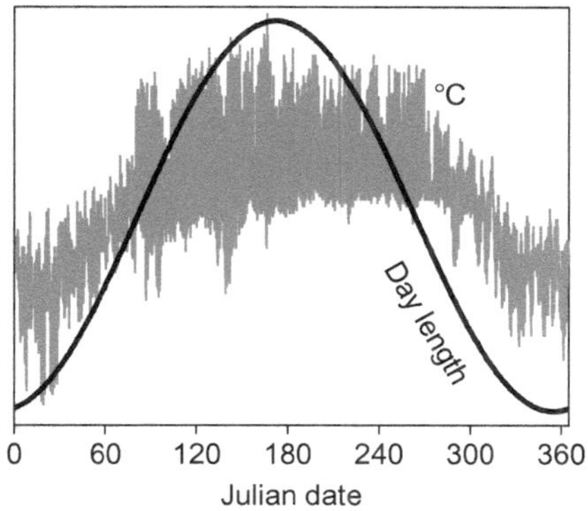

Fig. 1 Annual variation of day length and temperature. Temperature in a pitcher-plant leaf (°C, in *gray*) measured every 2 h for 1 year. For comparison, day length (*black line*) is shown for the equivalent date.

time, the annual fluctuation in day length increases in amplitude with increasing latitude, but not altitude (Bradshaw & Holzapfel, 2007). Hence, the timing of seasonal activities is dependent upon length of the local growing season, the local flux in day length, and the locality-appropriate response to day length. It is important to emphasize that the target of local selection is not day length itself. Rather, selection acts on the genetically determined response of individual organisms to length of day. Genetic variation underlying response to day length is tested in the crucible of the local seasonal environment, forming a feedback loop from the genome through photoperiodic response to concordance of resulting phenotypes with season-specific opportunities and exigencies. Seasonal selection then winnows out early or late phenotypes and alters allelic frequencies in genes programming photoperiodic response, thereby altering the timing of seasonal events. In sum, photoperiodic response represents the physiological connection between the genome and the seasonal phenotype and, as seasonal climate changes in space and time, so also do the observed photoperiodic responses of organisms that live in that climate.

1.3 Photoperiodism in *Wyeomyia smithii*

1.3.1 Photoperiodism and Phylogeography

Wyeomyia smithii overwinters as diapausing larvae in the evergreen leaves of the carnivorous purple pitcher plant. The robustness of its photoperiodic response and the consistency of its habitat over a broad geographic range result in a clear signal of seasonal and photoperiodic adaptation.

In order to understand the evolution and genetics of photoperiodism, it is important to use as an experimental organism one that is robustly photoperiodic in the first place. The pitcher-plant mosquito enters an hibernal diapause as larvae: Short days initiate and maintain diapause; long days promote continuous development or the resumption of development in diapausing larvae. The day length promoting 50% development and 50% diapause (the critical photoperiod) is the same for the initiation and termination of diapause in unchilled larvae (Bradshaw & Lounibos, 1972). For the initiation of diapause, insects pass through a "sensitive period" during which day length is physiologically interpreted as long or short, resulting in a diapause/no-diapause response (Beach, 1978; Saunders, 1981; Taylor, 1985). *Wyeomyia smithii* and some other insects are robustly photoperiodic while in diapause. For these insects the sensitive period is indefinitely long and the effect of manipulating different light:dark cycles can be assessed over weeks or months, instead of a few days, greatly enhancing their value in exploring photoperiodism.

Wyeomyia smithii oviposit into and complete their entire preadult development only within the water-filled leaves of the carnivorous pitcher plant *Sarracenia purpurea* (Smith, 1902), which is common in swamps and bogs from the Gulf of Mexico to Northern Canada and from sea level to high mountain seeps in the Appalachian Mountains. Throughout this range, *W. smithii* occupies a uniform microhabitat whose community composition remains highly consistent (Bradshaw, 1983; Bradshaw & Creelman, 1984; Buckley, Miller, Ellison, & Gotelli, 2003, 2010).

Dispersal of *W. smithii* in North America has been coincident with the historical biogeography of peatlands during and following retreat of the Laurentide Ice Sheet (Gorham, Lehman, Dyke, Janssens, & Dyke, 2007; Halsey, Vitt, & Gignac, 2000; Istock & Weisburg, 1987; Merz et al., 2013). Consistent with this distinction between the descendents of more ancient populations in the south and more recent populations in the north, isolation by distance is about six times greater among southern than northern populations (Armbruster, Bradshaw, & Holzapfel, 1998). Hence, the phylogeography of *W. smithii* presents diverse end points of evolutionary dispersal in space and time. Concomitantly, photoperiodic response has been exposed to diverse seasonal changes, both in situ and during postglacial dispersal. This history of populations over climatic gradients in space and time provides the necessary background for interpreting the evolutionary genetics of photoperiodism in *W. smithii.*

1.3.2 Geography of Photoperiodic Response

Geographically variable seasonal climates select for geographically variable photoperiodic responses. Critical photoperiod in *W. smithii* tracks the climatic gradient of North America more closely than any other known ecogeographic trait. Comparing the action spectra of light regulating diapause with light availability in pitcher-plant leaves during twilight defines a photoperiodic day and enables accurate prediction of the timing of diapause in nature from photoperiodic response determined under appropriately controlled conditions in the laboratory.

As one proceeds northward in latitude or upward in altitude, winters arrive earlier and springs arrive later in the year, shortening the length of the growing season. The optimal time to enter diapause is just over one full generation in advance of the onset of winter. The reason is that, at any given point in developmental time, an insect has either to continue developing or to enter diapause. If the insect continues to develop beyond its sensitive period, it must complete an additional full generation before its next sensitive period and be able to replace itself with at least one diapausing offspring in the overwintering population (Taylor, 1980). There are two important consequences of this reality. First, the diapause/no-diapause switch must be made when temperatures are declining. The optimal time to enter diapause then occurs when there remains more than the minimal number of day-degrees above threshold required to complete an entire generation. This go/no-go point can occur at a time of year when current environmental conditions are otherwise optimal for growth, development, and reproduction. This scenario underscores the importance of photoperiodism: Photoperiodically controlled regulation of diapause can override favorable temperature, humidity, and resources, ensuring that the individual will be in diapause when winter arrives.

Second, the earlier winter arrives, the earlier is the optimal time to enter diapause. Winter arrives earlier at higher than lower latitudes or altitudes. Hence, the optimal time to enter diapause should be earlier and at longer day lengths at higher than lower latitudes or altitudes. Consequently, the critical photoperiod-mediating diapause should increase with increasing latitude and altitude. The correlation between critical photoperiod and latitude has long been recognized (Danilevskii, 1965), but the relationship between altitude and latitude was first established in *W. smithii* (Bradshaw, 1976). Photoperiodic response curves mediating diapause in *W. smithii* show remarkably parallel sigmoid patterns across populations from the Gulf of Mexico to Canada and from sea level to over 1000 m in the Appalachians

(Bradshaw, Holzapfel, & Mathias, 2006). The photoperiodic response curves shift consistently from left to right with increasing latitude, with the exception of populations from the southern Appalachian Mountains at 35°N whose curves approximate those from low elevations at 40–41°N. These earlier studies, run at a constant temperature, were extended by using a sine-wave thermoperiod that varied from 13°C to 29°C each day and lagged the photoperiod by 3h as in nature (Lair, Bradshaw, & Holzapfel, 1997). Again, critical photoperiod among eight independently collected populations increased with latitude and altitude of population origin ($R^2=0.0.98$, $P<0.001$). In total, the regression of critical photoperiod on latitude and altitude has been repeated on four separate occasions using two different approaches and consistently yields $R^2 \geq 0.92$ (Bradshaw & Holzapfel, 2001). Critical photoperiod evolved 10 standard deviations across the range of populations sampled, indicating directional selection on a continental scale. By contrast, standard deviations of critical photoperiod did not vary with either latitude or altitude ($R^2=0.14$, $P=0.682$), indicating stabilizing selection on a local scale (Hard, Bradshaw, & Holzapfel, 1993; Lair et al., 1997).

The question remains as to whether critical photoperiods determined in the laboratory even under carefully controlled conditions are relevant to the timing of development in nature. To answer this question, Bradshaw and Phillips (1980) determined the astronomic units defining a photoperiodic "day" in nature by comparing the action spectra for photoperiodic response with the light-availability spectra within pitcher-plant leaves during twilight in nature. They determined that a day began at the onset of civil twilight in the dawn and extended through the end of civil twilight in the dusk. These results enabled using the critical photoperiods from Bradshaw, Quebodeaux, and Holzapfel (2003a), adjusted for twilight duration, to plot subjective day lengths as perceived by *W. smithii* on latitude and time of year, generating lines of iso-day lengths (Fig. 2). Actual time of entry into diapause in the field at four localities separated by more than 15° of latitude was accurately predicted by critical photoperiods determined in the laboratory.

2. GENETICS OF PHOTOPERIODIC RESPONSE

Photoperiodic response among populations "breeds true" under constant conditions in the laboratory and is highly heritable. Extant populations in nature harbor sufficient genetic variation in photoperiodic response to respond to rapidly changing seasonal environments. Photoperiodism is a

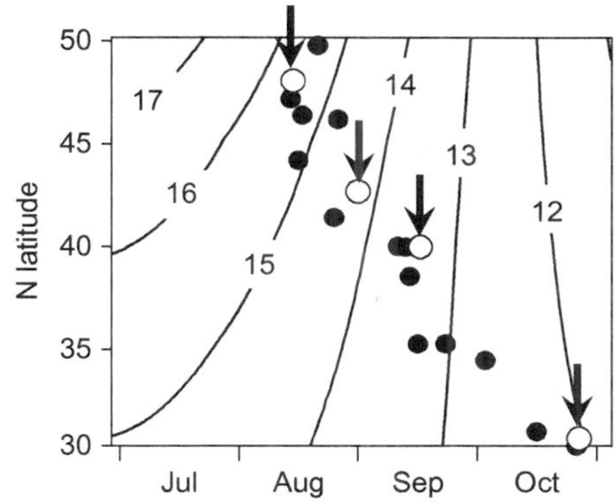

Fig. 2 Day length, latitude, and initiation of diapause. "Iso-day lengths" (sunrise–sunset with flanking civil twilights) vary with latitude and time of year. Critical photoperiods determined in computer-controlled environment rooms (*black circles*) are plotted according to latitude. The actual date of entry into diapause in the field is known for four localities (*open circles*). The *black arrows* indicate the date predicted from lab critical photoperiods for the initiation of diapause at the same four localities.

polymorphic trait, determined by many genes of varying effect. The additive effect of genes is immediately available to respond to selection; other genetic effects appear only in changing environments (genotype by environment interaction) or are hidden by interaction between alleles at the same locus (dominance relationships) or at different loci (gene–gene interaction or epistasis). The genes themselves remain elusive, but their combined effects and regions of the chromosome where they reside can be determined statistically using quantitative genetics. Evolution of photoperiodic response within and between populations of *W. smithii* has involved a complex web of both additive and nonadditive effects, resulting in unique genetic architectures underlying individual populations. There is no "wild-type" photoperiodic response in *W. smithii* and any evaluation of genetic effects must take the population-specific genetic background into account. Unraveling the genetics of the seasonal photoperiodic timing mechanism is going to be far more complex than that of the daily circadian clock.

2.1 Genetic Potential to Respond to Selection

Heritable genetic variation for photoperiodic response in *W. smithii* exists within and among populations. Heritabilities within populations are all nonzero, but increase dramatically in northern, postglacial populations.

Genetic variation in photoperiodic response among populations is readily evidenced by the fact that differences among populations persist in "common garden" experiments. In common garden experiments, populations are reared for two or more generations under the same conditions in the lab and then are exposed to identical experimental conditions. Since the environment is being held constant, differences that persist among populations are genetic in origin. In addition, like Danilevskii's (1965) pioneering experiments, hybrids between populations of *W. smithii* differing in photoperiodic response exhibit intermediate phenotypes (Bradshaw & Lounibos, 1977; Hard, Bradshaw, & Holzapfel, 1992; Hard et al., 1993; Lair et al., 1997).

The precise geographic cline in photoperiodic response and phenotypic variation within populations predicts that there should be substantial genetic variation in populations that is available to respond to selection in climates that change in space and time. Genetic potential to respond to selection depends on the heritability of a trait. The problem in estimating heritability for photoperiodic response is that this trait is typically assessed by the incidence of diapause in a cohort of individuals exposed to a given experimental regimen. Hence, critical photoperiod is described as a property of a group of organisms, not individual organisms. Because *W. smithii* responds to day length while in diapause, diapausing larvae can be exposed to short days that increment by 2–3 min per day as at mid-latitudes in nature. At some point, each individual interprets the day length as long and develops. The day an individual pupates is then scored as its individual critical photoperiod. Critical photoperiod is then directly expressed as a continuously distributed trait with mean and variance. Mean and standard errors of heritabilities are then estimated from the slope (*b*) from regression of offspring mean critical photoperiod on parent mean critical photoperiod across multiple parent–offspring cohorts (Bradshaw & Holzapfel, 1990). Critical photoperiods estimated from this approach include recognition of long day lengths and accumulation of photoperiodic information required to initiate development. Heritabilities of critical photoperiod ranged from 0.15 to 0.27 in three southern populations and 0.35 to 0.79 in three northern populations. Hence, extant populations harbor the genetic potential to respond to selection on photoperiodic response and this potential is higher in northern than southern populations.

2.2 Hidden Genetic Variation

Additive genetic variance is the primary, major source of phenotypic response to selection in large, outbreeding populations. When environmental conditions change, so also can the additive effects of genes (genotype by environment interaction or GxE). GxE can itself be altered by selection and can be a source of genetic variation contributing to adaptation. During migration and the founding of new isolated populations, population size is initially small and genetic drift becomes an important component of the genetic composition of populations. Isolation and drift can unmask recessive and epistatic alleles formerly hidden within ancestral populations. During population expansion following a founder event, selection assumes an ever-increasing role not only on phenotypes under selection but also on novel combinations of genes contributing to evolving phenotypes (Fisher, 1958; Gibson & Dworkin, 2004; Mayr, 1954; Paaby & Gibson, 2016; Paaby & Rockman, 2014; Phillips, 2008; Wallace, 1991; Wright, 1968, 1977; Yadav, Dhole, & Sinha, 2016).

2.2.1 Genotype by Environment Interaction

Differential response to day length represents a form of developmental phenotypic plasticity (phenotypic variation of an individual in different environments) and the heritability of photoperiodic response constitutes genetic variation in phenotypic plasticity, i.e., genotype by environment interaction (GxE). Selection on critical photoperiod exposes responses to exotic day lengths not ordinarily seen by populations in nature, thereby revealing GxE as a potential source of genetic variation in changing seasonal environments.

The first "hiding place" of genetic variation lies in the interaction between genotype and environment (GxE). Diapause in *W. smithii* represents a form of phenotypic plasticity: depending on the photic environment (day length), individuals may or may not enter diapause and critical photoperiod is a heritable trait, i.e., diapause is determined by interaction between the genotype and the environment (Chevin & Lande, 2015). One of the hallmarks of GxE's contribution to hidden genetic variation is a phenotype that is differentially expressed under conditions to which the population is never exposed. After imposition of truncation selection for long and short critical photoperiods in three subpopulations from New Jersey (Bradshaw, Quebodeaux, & Holzapfel, 2003b), developmental response to exotic short days never experienced in nature is enhanced in the short-selected lines and reduced in the long-selected lines. Even more unexpected is marked reduction in diapause at exotic long day lengths in both the short- and long-selected lines. Selection has clearly revealed hidden genetic variation for diapause response to exotic day lengths. Hence, GxE can increase the ability of populations to respond to changing seasonal climates, including the advance and regression of glaciers or recent rapid climate change.

2.2.2 Dominance and Epistasis

Interaction between alleles both within and among populations provides novel genetic variation to explain complex genetic architectures underlying photoperiodism and its evolutionary flexibility.

The second hiding place of genetic variation lies in allelic and genic interactions, i.e., dominance and epistasis that are revealed by independent assortment and recombination in hybrid phenotypes. Additive and nonadditive genetic effects on differences in critical photoperiod among latitudes, longitudes, and altitudes are shown in Table 1 based on 14-generation line crosses (Lair et al., 1997). Expected generation means were estimated from F-∞ parameterization as in Hayman (1958, 1960a, 1960b) and Mather and Jinks (1982, Chs. 5, 10). Models of genetic effects were tested, sequentially, for adequacy of additive, additive-dominance, additive-dominance-maternal, and additive-dominance-maternal-digenic epistasis models. Rejection of the most inclusive model indicated significant higher order epistasis and/or linkage effects.

Several patterns emerge. First, there are always strong additive effects contributing to the genetic differences in critical photoperiod among populations. Second, either dominance or maternal dominance effects

Table 1 Differences in Genetic Effects Among Populations

Cross	Add	Dom	Amat	Dmat	AxA	AxD	DxD	ADME
FLxAL	***	***	*	***	***	°	**	°
FLxMt	***	°	°	**	°	***	*	°
NCxMt	***	***	°	°	°	°	***	°
MtxME	***	°	***	*	***	°	°	***
NJxON	***	°	°	**	—	—	—	—
WIxME	***	***	°	°	—	—	—	—
MExON	***	**	**	**	°	°	***	°

Significance of genetic effects given by: °$P>0.05$; *$P<0.05$; **$P<0.01$; ***$P<0.001$ after row-wide sequential Bonferroni correction. Add, additive; ADME, test for adequacy of the combined additive-dominance-maternal model; Amat, additive maternal; AxA, additive by additive digenic epistasis; Ax-D, additive by dominance digenic epistasis; Dmat, dominance maternal; Dom, dominance; DxD, dominance by dominance digenic epistasis. A dash (—) indicates no tests for digenic epistasis were made because the additive-dominance-maternal model was adequate to explain genetic differences in critical photoperiod between parent populations. Mt indicates a population from the southern Appalachian Mountains.

Summarized from Lair, K. P., Bradshaw, W. E., & Holzapfel, C. M. (1997). Evolutionary divergence of the genetic architecture underlying photoperiodism in the pitcher-plant mosquito, *Wyeomyia smithii*. *Genetics*, *147*, 1873–1883.

contributed to genetic differences in critical photoperiod among populations. Third, digenic epistasis made significant contribution to genetic differences in critical photoperiod in five of the seven crosses; the two exceptions both occurred in crosses between two populations within the northern clade (WIxME and NJxON) and not between the cross within the southern clade (FLxAL) that was separated by less geographic and phenotypic distance. Third, there was no consistent pattern of digenic epistasis among crosses, supporting Hard et al.'s (1992, 1993) conclusion that different genetic architectures contributed to the evolutionary divergence of critical photoperiod among populations. What Table 1 shows is that rampant and diverse hidden genetic variation for photoperiodic response exists without any apparent pattern throughout the range of *W. smithii*. Evolution of photoperiodic response in *W. smithii* is founded on a rugged genetic landscape of allelic and genic interactions.

During dispersal of *W. smithii* in North America, the additive genetic variance for critical photoperiod has increased exponentially; over the same range, average heterozygosity of individual protein-coding loci has declined, as would be expected from sequential founder events following recession of the Laurentide Ice Sheet 20,000 years ago (Armbruster et al., 1998). Based on a consistent involvement of epistasis in south to north differentiation of postglacial populations, Hard et al. (1993) attributed this increase in additive genetic variance to a release of additive from nonadditive genetic variance during isolation and drift following successive founder events.

For there to be a release of additive from nonadditive genetic variance during the differentiation of populations, there has to be nonadditive genetic variation within populations in the first place. To test this proposition, Bradshaw, Haggerty, and Holzapfel (2005) tested for nonadditive genetic variance in a single New Jersey meta-population (40°N). They collected from three interconnected subpopulations within a 200 m radius in this large population and, within each subpopulation, imposed selection for divergent critical photoperiod for 13 generations. After selection, they performed a 14-generation line cross within each. In each case, the joint scaling test rejected an additive-dominance-maternal model and the signature of digenic epistasis was unique to each subpopulation. These results showed first that nonadditive genetic variance for photoperiodic response existed in all three of these sites, consistent with the concept of there being nonadditive genetic variance available within populations to be converted to additive genetic variance. These results also showed that there was spatial genetic structure within the New Jersey meta-population over a range of just 200 m, that early events in their responses to selection sent them down different genetic trajectories, or a combination of both.

Short-term selection within a single meta-population of *W. smithii* revealed sufficient nonadditive genetic variance to explain the release of additive from nonadditive variance during their postglacial dispersal. More importantly, there exists within a single meta-population of *W. smithii* the full range of potential allelic and genic interactions observed across the entire species from Florida to Canada shown in Table 1.

2.3 Mapping Genotype to Phenotype

The number of genes and position of genes within the genome that contribute to the evolution of photoperiodic time measurement can be estimated from line crosses and quantitative trait loci (QTL), respectively. The latter approach reveals polymorphism in QTL within and between populations and confirms that the perception of genes contributing to photoperiodic response is crucially dependent on genetic background.

While a great deal is known about the genetics of the daily circadian clock in insects, including at the molecular level within species and between orders of insects (Tormey et al., 2015), the actual genes involved in the seasonal photoperiodic timer, apart from being *somehow* connected to the circadian clock (Section 4), remain elusive. The number of genes can be estimated from line cross means and variances via the Castle-Wright-Lande model that estimates the minimum number of "effective factors" contributing to genetic differences among populations or lines (Lynch & Walsh, 1998, p. 233; Roff, 1997, p. 11). For photoperiodic response, these estimates range from 0 to 19, and increase with phenotypic distance between the parent populations (Hard et al., 1992; Lair et al., 1997).

An alternate approach is to use QTL mapping to estimate the number and effect of regions in the genome containing the genes or groups of genes contributing to genetic differences in critical photoperiod. In the first QTL for photoperiodic response in any animal, Mathias, Jacky, Bradshaw, and Holzapfel (2007) made an F_2 hybrid cross between geographic extremes of *W. smithii*, a Florida ♀ (31°N) × an Alberta ♂ (57°N). The critical photoperiods of both the parents and F_2's were determined with incremental day lengths and genotyped using AFLP and gene-based markers. The QTL map was based on 45 marker loci spanning 287 cM from 264 F_2 individuals. *Wyeomyia smithii*, like other culicine mosquitoes, has three chromosomes (Moeur & Istock, 1982). Composite interval mapping (CIM) revealed nine

QTL: two on the first chromosome bearing the sex locus, six on the second chromosome, and one on the third, accounting for an estimated 62% of the phenotypic variance. The latter portion of the second chromosome includes an overlapping region of QTL for photoperiodic response as well as stage of diapause. Also included in this region are epistatic interactions between one of the markers and five other markers within this region, including two ecdysteroid-activated genes, *Pep* and *Impl3*. This region of the second chromosome is then a candidate "hot spot" for genes integrating continuous development, photoperiodic response, diapause, and metabolism in *W. smithii* (Mathias et al., 2007).

Individual QTL represent a region of the genome and may include multiple genes. In addition, resolution of QTL is limited by the number of markers, the number of individuals phenotyped and genotyped, and the number of recombination events preceding the genotyping generation (Beavis, 1998; Mackay, Stone, & Ayroles, 2009; Phillips, 2005; Xu, 2003). In a separate series of crosses involving entirely different localities, Bradshaw, Emerson, Catchen, Cresko, and Holzapfel (2012) and Bradshaw, Emerson, and Holzapfel (2012) examined QTL for photoperiodic response among three populations based on single-pair matings of *W. smithii*: New Jersey ♂ × Maine ♀ (NJ × ME) separated by ~13,000 years of evolutionary time, and reciprocal single-pair matings of Alabama × Maine (AL × ME, ME × AL) separated by ~200,000 years of evolutionary time (Merz et al., 2013). RAD markers greatly increased the resolution of QTL with an average of 220 markers and 660 cM per cross.

CIM of photoperiodic response between the New Jersey and Maine populations revealed two QTL, one on the second and one on the third chromosome. CIM of the AL♂ × ME♀ revealed three QTL, all on the first chromosome. Finally, CIM of the reciprocal ME × AL revealed two QTL, one on the first and one on the second chromosome. There was no overlap in QTL among the three maps. Both of the QTL in NJ × ME showed dominance deviation toward the southern parent; all of the QTL in AL × ME showed dominance deviation toward the northern parent; and the one QTL in the reciprocal ME × AL showing significant dominance deviation was toward the northern parent. Hence, there was no consistent position of QTL or sign of additive or dominance effects among the three crosses. Reciprocal crosses between different parents within the same population indicated as much variation in regions of the genome contributing to differences in photoperiodic response as among all crosses separated over wide geographic space and evolutionary time. Perception of regions of the genome, much less individual genes contributing to the evolution of photoperiodic response, depends crucially on the genetic background within individual populations.

2.4 Lessons From Quantitative Genetic Approaches in *Wyeomyia smithii*

Line crosses and QTL mapping of genomic regions contributing to divergence of photoperiodic response reveal complex underlying genetic architectures within populations, between nearby populations, and among populations that are more remote in space and time. In a sense, there is no standard genetic background or unique "wild type." Any evaluation of standing genetic variation within a population, genetic structure among nearby subpopulations, or potential genetic processes underlying current patterns of phenotypic variation or even phenotypic similarities has to take genetic background into account.

Both line-cross analyses and QTL mapping have revealed a complex genetic architecture underlying genetic variation within local populations as well among populations having dispersed over the climatic gradient of North America. Genotype by environment interaction, maternal effects, dominance, and epistasis provide hidden sources of genetic variation that potentially can be recruited during dispersal across climatic zones or within local populations during periods of climate change. Most importantly, convergent phenotypes can have divergent underlying genetic architectures and even local populations contain additive and nonadditive genetic variation sufficient to generate novel genetic architectures in descendent populations of a dispersing species.

3. PHOTOPERIODISM AND CLIMATE CHANGE

Recent rapid climate change has been imposing primarily seasonal, not thermal selection on the biotic world. Response to this selection has resulted in the expansion of animal activities during progressively longer growing seasons and in a genetic (evolutionary) shift in photoperiodic response toward more southern-like critical photoperiods.

The last five decades have brought a challenge to research scientists: Earth has been warming at an accelerated rate, unprecedented in the last millennium (Stott et al., 2000). Polar and temperate winters have become milder and shorter, and growing seasons have become longer. The abiotic effects of global warming have resulted in melting glaciers and rising sea levels (IPCC, 2007; Karl & Trenberth, 2003), but animals are not glaciers and the main effects on temperate and polar biota have been new opportunities resulting from the opening of new climatic zones poleward and from longer growing seasons without imposing appreciable heat stress (Bradshaw & Holzapfel, 2010b, 2010c). The result is that animals have been expanding their ranges poleward, and exploiting longer growing seasons by increasing time spent growing, developing, and reproducing, and by migrating earlier in the spring and later in the fall. In short, animals have been entering new photic zones and experiencing selection on the timing of seasonal activities in their life histories (Hughes, 2000; Parmesan & Yohe, 2003; Peñuelas & Filella, 2001; Root et al., 2003). Initially, these patterns could be ascribed to phenotypic plasticity, but genetic shifts in photoperiodic response in *Wyeomyia smithii* provided the first evidence that recent rapid climate change has actually been driving genetic change (evolution) in natural populations (Bradshaw & Holzapfel, 2001). Over a 24-year period, there has been a genetic shift toward shorter, more southern-like critical photoperiods. This shift is detectable after only 5 years, representing evolution at break-neck speed in nature. The shift is also more pronounced in the north than in the south, reflecting both the higher additive genetic variance for photoperiodic response in northern populations and the trend of climate change to be faster in the north than in the south. Evolutionary responses were subsequently confirmed in plants, other insects, birds, and mammals (Bradshaw & Holzapfel, 2006, 2008, 2010b; Franks, Sim, & Weis, 2007; Gomi, Nagasaka, Fukuda, & Hagihara, 2007; Urbansky et al., 2012), all reflecting seasonal, not thermal adaptation.

Definitive evidence of the relative importance of photic vs thermal adaptation imposed by rapid climate change was determined experimentally by Bradshaw, Zani, and Holzapfel (2004). They programmed year-long climates in computer-controlled environmental rooms capable of reproducing daily and annual changes in light, temperature, and humidity from the tropics to polar regions of Earth. As an index of fitness, they used the year-long cohort replacement rate, i.e., net performance integrated through all four seasons of *W. smithii* reared in the leaves of intact pitcher plants and fed freeze-dried Drosophila to mimic the prey-capture behavior of plants in nature. As a baseline they determined fitness of four northern populations in a natural 50° N thermal and photic year throughout all four seasons (Fig. 3). They then programmed the same northern photic year but a mid-latitude thermal year equivalent to 180 years of global warming at its present rate through all four seasons. Fitness actually increased by 47%—warmer was better. When they

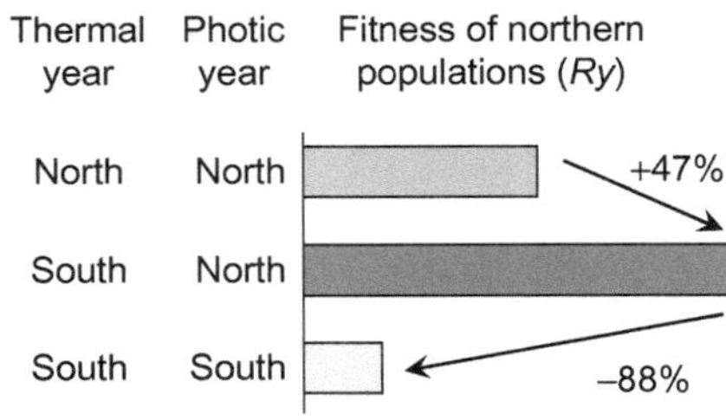

Fig. 3 Photoperiodism, temperature, and fitness. Year-long climates were programmed in computer-controlled environment rooms, replicating the daily and annual changes in light, temperature, and humidity in nature. Fitness (*Ry*) was determined for four northern populations under three different conditions: (1) northern thermal and photic years (*light gray*), (2) southern thermal and northern photic years (*dark gray*), and (3) southern thermal and photic years (*white*). *Ry* represents the year-long cohort replacement rate integrated over all four seasons of *Wyeomyia smithii* living in the leaves of intact pitcher plants.

then programmed the same benign mid-latitude thermal year, but also programmed the mid-latitude photic year, fitness declined by 88%. Hence, the immediate response to climatically incorrect photic cues resulted in a path to rapid extinction. Thermal adaptation does occur across latitudinal and altitudinal gradients over long timescales, but photic responses track geography more closely than thermal tolerance or thermal optima (Bradshaw & Holzapfel, 2008; Zani, Cohnstaedt, Corbin, Bradshaw, & Holzapfel, 2005). The important conclusion is that the immediate evolutionary response to rapid climate change in the biological world involves seasonal (photoperiodic) adaptation, with thermal adaptation occurring over a longer time span in already established populations.

4. PHOTOPERIODISM AND CIRCADIAN RHYTHMICITY

The evolutionary response of photoperiodism in *W. smithii* during both millennial and recent climate change offers decades if insight into the evolutionary and functional connection between daily and seasonal timekeeping. Both the daily circadian clock and the seasonal photoperiodic timer use the input of light, but serve different functional roles in biological timekeeping. Experimentation and intense discourse surrounding their relationship has persisted for over 80 years. *Wyeomyia smithii* addresses this discourse from an evolutionary perspective and this perspective is used to formulate a model reconciling more recent molecular experiments and the independent evolution of the clock and timer in *W. smithii*.

4.1 Functionality, Definitions, and the Evolutionary Perspective

There are two great rhythms of the biosphere: a daily, 24-h rhythm created by the rotation of the earth about its axis and a seasonal rhythm created by the rotation of the earth about the sun. Temporal organization of daily biochemistry, physiology, and activities is orchestrated by an internal, self-sustaining circadian clock. Temporal organization of seasonal biochemistry, physiology, development, and activities is orchestrated by day length. Concordance with both the daily cycle of dawn and dusk and the seasonal cycle of day length is crucial for the maintenance of fitness in *Wyeomyia smithii* (Bradshaw et al., 2004; Emerson, Bradshaw, & Holzapfel, 2007). The circadian clock and the photoperiodic timer are functionally distinct (Bradshaw & Holzapfel, 2010a; Danks, 2005; Hut, Paolucci, Dor, Kyriacou, & Daan, 2013):

> *Circadian rhythms entrain to dawn and dusk on a continuous basis while photoperiodism acts as a go/no-go seasonal switch that commits an animal to migration, dormancy, development or reproduction that may be separated from the present environment in time or space by months or thousands of kilometers... Entrainment of the circadian clock resets on a day-to-day basis ... while a photoperiodic response, once executed, is irreversible within a seasonal context or even within the lifetime of an individual ... Finally, the circadian clock does not "count" light:dark cycles. The photoperiodic counter both counts and accumulates light: dark cycles that the photoperiodic timer has interpreted as long or short and then triggers the corresponding physiological response when some threshold number of inductive cycles has been exceeded.*
>
> ***Bradshaw and Holzapfel (2010a, p. 156)***

Precise definitions are important in this discourse. Confucius called it the rectification of names: "That is, things in actual fact should be made to accord with the implication attached to them by names" (Fung, 1960, p. 41). For example, the often used term "photoperiodic clock" implies a priori unity of photoperiodism and circadian rhythmicity. Hence, the discussion below will make the consistent distinction, unless in a direct quote from the literature, between the daily circadian clock (the clock) and the seasonal photoperiodic timer (the timer) and between "photoperiod" (day length or light portion of the daily cycle) and "photoperiodism" (ability to use day or night length to regulate some biological, generally seasonal process).

Biological timekeeping in *W. smithii* has been interpreted in the light of geographic variation that represents diverse end points of evolution in space and time. The major perspective to keep in mind is that regardless of variable seasonal selection on photoperiodic response, organisms on Earth live in a constant 24-h world through evolutionary time. A major theme of this discourse revolves around the question: "If the circadian clock forms the basis of the photoperiodic timer, how can the clock exposed to an invariant 24-hour world maintain precise integration of daily events while the timer maintains genetic and evolutionary flexibility through highly variable space and time?"

4.2 Daily vs Seasonal Timing

Classical physiological experiments show that *Wyeomyia smithii* assesses the length of day, not night, and that tests for circadian rhythmicity as an integral component of photoperiodism yield negative or contrary results within and among populations. Unique comparisons between lowland and mountain populations of *W. smithii* at the same latitude (hence, same photic environment, but different seasonal environments) show independent evolution of the photoperiodic timer and the circadian clock, the latter measured by overt circadian behavior and by the rhythmic expression of the core circadian clock gene *period*. However, accumulating evidence from other insects supports the concept that the clock and timer are "somehow related." Herein, we propose a model based on gene pleiotropy that reconciles these seemingly disparate results and viewpoints.

4.2.1 Bünning's Proposition Then and Subsequently

Because both circadian rhythmicity and photoperiodism are involved in biological timing and both rely on the input of light, there has been avid conjecture, discussion, and experimentation revolving around the mechanistic connection between them ever since Bünning (1936, p. 590) proposed that the circadian clock formed the *Grundlage* of photoperiodic time measurement: A great deal hangs on the meaning of *Grundlage*, which can be translated as groundwork, foundation, base, rudiment, matrix, or basement (De Vries, 1959). In an English edition, Bünning (1973) is not clear what he meant by *Grundlage*, but, based on the observation that critical photoperiod is often independent of temperature, "This [independence] suggests the assumption that here, too, [photoperiodic] time is measured by means of the circadian clock" (p. 203) and, in retrospect (p. 223), "it is no longer surprising to us that certain developmental processes may or may not be coupled to the clock." Bünning's proposition has historically elicited voluminous and creative physiological and genetic experiments with an increasing appreciation of an evolutionary context. Investigations with *Wyeomyia smithii* emphasize evolutionary and physiological genetics, raise fundamental questions about the interpretation of data and the formulation of hypotheses, and conclude that Bünning's "coupling," if carefully defined, may achieve reality by invoking gene pleiotropy.

4.2.2 How *W. smithii* Assesses Day Length

Photoperiodic timing in *Wyeomyia smithii* relies on the day length and not night length, and requires the repeated input of external light rather than an internal circadian clock to count photoperiodic inductive cycles. However, comparisons among populations of this single species reveal a more complex evolution of photoperiodism than would be perceived from consideration of a single population.

There are two fundamental questions relating to photoperiodism in any organism. First, does the photoperiodic timer assess the length of day or the length of night (Saunders, 2013) and, second, does the accumulation of photoperiodic information (the photoperiodic counter) continue under constant conditions (darkness) or require the repeated input of light? In diapausing populations of *W. smithii* from the Gulf of Mexico to Canada, day lengths longer than the critical photoperiod invariably induce development, regardless of night length (Bradshaw, Holzapfel, & Davison, 1998; Emerson, Bradshaw, & Holzapfel, 2009; Emerson, Dake, Bradshaw, & Holzapfel, 2009). Photoperiodic response in *W. smithii* relies on day length, not night length.

In an experiment with eight populations from the Gulf of Mexico to Canada, diapausing larvae were exposed to a diapause-terminating long day every 24, 48, or 72 h (Emerson, Letaw, Bradshaw, & Holzapfel, 2008). If the counter were continuing to run in the longer nights of the 48- and 72-h cycles, each of the longer cycles should count twice or three times as many long days as a 24-h cycle, respectively. In fact, all three cycle durations counted an equivalent number of long days. In addition, when Bradshaw et al. (2003b) exposed 14 populations from the Gulf Coast to Canada to day lengths from 10 to 18 h with at total cycle of 24 or 72 h, the critical photoperiods were the same for both treatments. These results show that the photoperiodic counter in *W. smithii* relies on the repeated input of light to accumulate photoperiodic information, but the photoperiodic timer does not lose track of day length during long nights.

Pittendrigh and Minis (1964) proposed that light played two roles: setting the phase of the sensitivity rhythm and triggering a long-day response if light occurred during the sensitive phase of the rhythm (the external coincidence model). The model predicts that in an otherwise diapause-maintaining short day regimen, there should be two peaks of diapause-terminating sensitivity to light pulses during the long night in *W. smithii*: an early night "A" peak in effect simulating a delayed dusk after the main short day, and a late night "B" peak simulating an advanced dawn before the main short day. With increasing critical photoperiod, the A peak should occur progressively later in the night and the B peak progressively earlier (Saunders & Bertossa, 2011).

This effect of light pulses was tested by exposing six populations of diapausing *W. smithii* from the Gulf of Mexico to Canada to a short day and a long night with both T = 24 (L:D = 10:14) and T = 48 (L:D = 10:38) interrupted by 1-h light pulses every hour (Bradshaw et al., 1998). As expected, southern populations showed both A and B peaks, but the peaks did not change position with either T = 24 or T = 48. Furthermore, the A peak persisted at all latitudes; the B peak declined with increasing latitude. When T = 48, there appears an unpredicted third peak between the A and B peaks in the southern populations only.

While these results are consistent with Pittendrigh and Minis' (1964) concept of the dual action of light (setting the clock and triggering the long-day response), it is also clear that photoperiodic response in *W. smithii* is more complex than envisioned by this model. These results also demonstrate the importance of the evolutionary perspective: had only a mid-latitude population been investigated, the complexities of the disappearing B peak in the north and the appearance of the C peak in the south would not have been revealed.

4.2.3 Resonance

As evidenced by resonance (tracking external light:dark cycles) experiments, variation in expression of the daily circadian clock is not a driving force in the evolution of the seasonal photoperiodic timer in *Wyeomyia smithii*.

Short days and long nights should, at first glance, maintain diapause regardless of night length. However, this condition exists only if the circadian clock can track or resonate with the total light:dark (L:D) cycle. Conceptually, a circadian-driven light-sensitivity rhythm should resonate with cycles of ~24, 48, or 72 h, i.e., short days and long nights should maintain diapause. With cycles of ~36 or 60 h that depart radically from the normal 24-h period of the environment, the circadian clock should fail to synchronize with the L:D cycle and the sensitive phase of the cycle should drift into the short day, resulting in a discordant long-day response. When exposed to short days and increasing night lengths (in different experiments) to create cycles of T = 10:14, 10:16, 10:18, ... 10:62, a circadian-driven photoperiodic timer should encounter resonant cycles that maintain diapause and discordant cycles that stimulate development. Hence, a plot of percent development as a function of T = 10 + D should oscillate between diapause and development as the underlying circadian sensitivity rhythm resonates with or drifts through the external L:D cycle. Such experiments are known as resonance, T, or Nanda–Hamner experiments (Pittendrigh, 1981; Saunders, 2002, pp. 351–358).

Response by *Wyeomyia smithii* to resonance experiments shows that there is the expected rhythmic response in Gulf and Carolina lowland populations. The period of this rhythm remains constant at ~21 h throughout the latitudinal and altitudinal range of *W. smithii*, but the amplitude of this rhythm varies from robust to nonexistent (Bradshaw et al., 2003a, 2003b; Wegis, Bradshaw, Davison, & Holzapfel, 1987). Amplitude of the rhythmic response to resonance experiments is correlated with altitude, but not with latitude and not with critical photoperiod, showing that the photoperiodic timing of seasonal events in *Wyeomyia smithii* can evolve independently of the daily circadian clock (Bradshaw et al., 2006).

The declining amplitude of resonance rhythmicity might be explained by a rapidly damping circadian oscillator (Saunders, 2009, 2016). If, instead of exposing diapausing larvae to a resonance experiment with short days, they are exposed to a resonance experiment with diapause-terminating long days, >92% of all larvae develop and eventually pupate (Emerson, Dake, et al., 2009). Development time measured as days to pupation increases with T = L + D, but, in some populations, there is rhythmic variation about the regression of development time on T. Notably, low-elevation populations at 35°N have a robust short critical photoperiod, a robust resonance when L = 10 h, and a robust rhythmic number of days to pupation when L = 18 h. By contrast high-elevation populations at the same latitude have a long critical photoperiod, a totally arrhythmic, zero resonance response, but a robust rhythmic number of days to pupation when L = 18. All experiments were run at the same temperature. The rhythmic days to pupation do not differ in period or amplitude between the low- and high-elevation populations. As evidenced by resonance of development time with varying external light:dark cycles, a robust circadian oscillator persists in populations at both elevations; consequently, a damped circadian oscillator cannot be invoked to explain the absence of a rhythmic response to resonance experiments in *W. smithii*.

Pittendrigh and Minis recognized the power of an indefinitely long sensitive period when photoperiod controls the maintenance and termination of diapause as they observed in the pink bollworm: "In general, resonance experiments are (1) always powerful when a positive resonance effect is found; but (2) powerful when one is missing only if the system remains for a sufficiently long time in the inducible state for the resonance effect to occur" (Pittendrigh & Minis, 1971, p. 243). The sensitive period in *W. smithii* persists throughout diapause for as long as year and the resonance experiments were of 8 weeks duration. The 8 weeks provided ample time for >95% development to long-day control 18:54=L:D cycles, and therefore any positive resonance effect to occur. The absence of a positive resonance effect in independent mountain populations is not an anomaly, but instead constitutes the powerful experiment called for by Pittendrigh and Minis and demonstrates that something other than a simple concept of the daily circadian clock being an integral component of photoperiodism is driving evolution of the seasonal photoperiodic timer in *Wyeomyia smithii*.

4.3 Molecular Connections and Pleiotropy

Increasing evidence from mutations and RNA interference of circadian clock genes supportPittendrigh's (1972, p. 2734) concluding assertion that "circadian rhythmicity is often *somehow involved* in the physiology of photoperiodic induction" (emphasis ours). Over four subsequent decades of physiological, genetic, and molecular research has not substantially improved on the "somehow involved." Below, we present a model that reconciles the diverse evolutionary, physiological, functional, and molecular viewpoints of how the circadian clock is "somehow involved" in photoperiodism. This reconciliation depends on the recognition that evolution, physiology, function, and mechanism are, in fact, different entities and as such may be subject to different selective pressures, regardless of their interrelationships.

The circadian clock and the photoperiodic timer, while both cued by light, regulate different functions, daily and seasonal timing, respectively (Bradshaw & Holzapfel, 2010a; Danks, 2005). As discussed in Section 4.1, whereas the earth's rotation about its axis imposes a 24-h world regardless of geography or climate, the rotation of the earth about the sun generates seasonal climates that vary directly with geography (MacArthur, 1972) and that have profound effect on photoperiodism (Bradshaw &

Holzapfel, 2006, 2007, 2010b; Danilevskii, 1965; Danks, 2005; Hut et al., 2013). There is now accumulating evidence from molecular data supporting the proposition that the circadian clock is "somehow involved" in the mechanism of photoperiodic time measurement (Doležel, 2015; Goto, 2013; Hut et al., 2013; Koštál, 2011; Meuti & Denlinger, 2013; Numata, Mihazaki, & Ikeno, 2015; Omura, Numata, & Goto, 2016). What remains to be resolved 80 years after Bünning's (1936) proposal and Pittendrigh's refinements of the concept (Pittendrigh, 1966, 1972; Pittendrigh & Minis, 1964) assertion is still what "somehow involved" means in terms of mechanism at the genetic and genomic levels.

Although many experiments spanning decades of research with *Wyeomyia smithii* show the signature of independent evolution of the circadian clock and the photoperiodic timer, the most direct evidence for this conclusion comes from comparing photoperiodic and circadian properties over an elevation gradient at the same latitude. In this comparison, both high- and low-elevation populations experience the exact same annual change in day length, each day of the year. However, lower elevations have earlier springs, later onsets of winter, and longer growing seasons than higher elevations. Hence, there is selection for more northern-like photoperiodic responses at high mountain altitudes as compared with low elevation. *Wyeomyia smithii* from 35°N in North Carolina illustrates just this expected result (Bradshaw, 1976; Bradshaw et al., 2003a, 2003b). The question then remains as to what are the properties of the circadian clock over this same elevation gradient?

At the physiological level, diapausing *W. smithii* from high and low elevations exposed to a diapause-terminating long day followed by night lengths varying from 6 to 54 h ($T = L + D = 24$–72 h) in 26 different experiments exhibited rhythmic development time with an average period of 23 h (Section 4.2.3). However, neither period nor amplitude differed between the mountain and lowland populations, indicating that the overt properties of the circadian clock were the same regardless of differences in seasonality and photoperiodic response.

At the molecular level (Fig. 4), rhythmic expression of the core circadian clock gene *period* in constant dark did not differ between the mountain and lowland populations in period, amplitude, or degree of damping.

These experiments show that among populations, when the photic environment is held constant while the seasonal environment is varied, physiological and molecular properties of the circadian clock remain invariant and constant while photoperiodic response follows the seasonal environment.

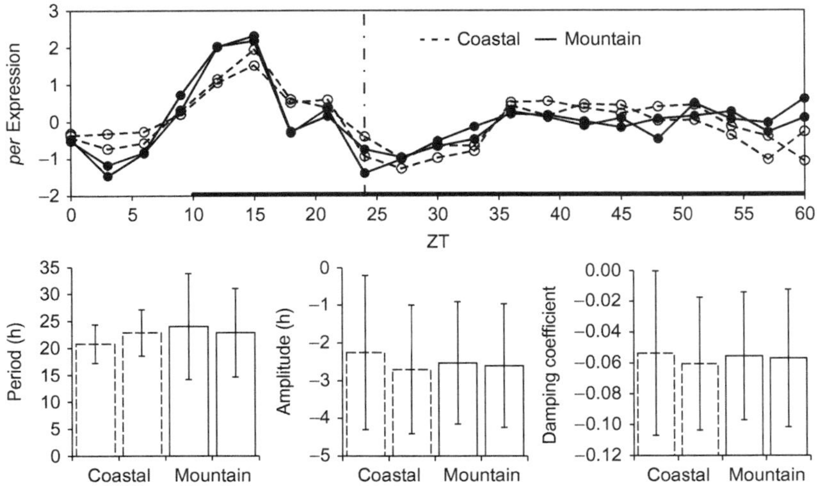

Fig. 4 Rhythmic properties of period. The *top plot* shows *per* expression in diapausing *Wyeomyia smithii* larvae as log_2 fold variation about mean expression for each population. Two populations each were from either the coastal region or the Appalachian Mountains in North Carolina at the same latitude, 35°N. In nature, these mosquitoes and localities encounter the same year-round day lengths but very different seasonal environments. When the photic environment is held constant, but the seasonal environment is varied, molecular properties of the circadian clock remain invariant while the photoperiodic response follows the seasonal environment.

Within a single population, the sign of a genetic correlation between critical photoperiod and rhythmic response to resonance experiments can be reversed by antagonistic selection even more readily in *W. smithii* (Bradshaw et al., 2012) than in *Drosophila littoralis* (Lankinen & Forsman, 2006). Therefore, these two great biological timing mechanisms are clearly capable of independent response to selection and, hence, independent evolution. This conclusion returns us to the conundrum: how does the circadian clock maintain close control of daily functions in a 24-h world if it is tightly coupled to a genetically flexible photoperiodic timer finely attuned to geographically local seasonal environments? The answer depends on how one asks the question. If one asks, "How does the circadian clock control photoperiodism" one is already assuming a priori that the circadian clock controls photoperiodism. Koštál (2011) and Doležel (2015) pose the more neutral question: "What is the relationship between the circadian clock and the photoperiodic timer—is it one of unity, cooperation, or independence?" Given the evolutionary independence of the clock and timer in *W. smithii*, the question is more appropriately asked: "How can the photoperiodic timer connect with the circadian clock?" In other words, how can a variable photoperiodic timer co-opt the time cues provided by the circadian clock without disrupting daily temporal organization? Emerson, Bradshaw, et al. (2009) and Bradshaw and Holzapfel (2010a) contrasted the circadian clock acting as a unit or module (modular pleiotropy) with individual circadian clock genes acting individually, and incidentally of their circadian function on photoperiodic timing (gene pleiotropy). Tormey et al. (2015) and Denlinger, Hahn, Merlin, Holzapfel, and Bradshaw (2017) proposed a variant of the gene pleiotropy model (Fig. 5) that falls between Koŝtál's (2011) and Doležel's (2015) concepts of independence and cooperation. Basically, it posits that the circadian clock must communicate with the rest of the organism and it is most likely to do so through the transcription factors and transcription regulators involved in the core transcription–translation feedback loop. All it takes is for the photoperiodic timer to cue in on, or co-opt one or a few of the output proteins of these genes to provide a time reference point. Herein, we further refine this proposition with the "commensal model of biological timekeeping." In this model, the relationship is one of commensalism (one entity, the photoperiodic timer, benefits without cost or benefit to the other entity, the circadian clock), not parasitism or mutualism. Evolution of photoperiodic response can then be the result of the photoperiodic timer co-opting different titers of the same

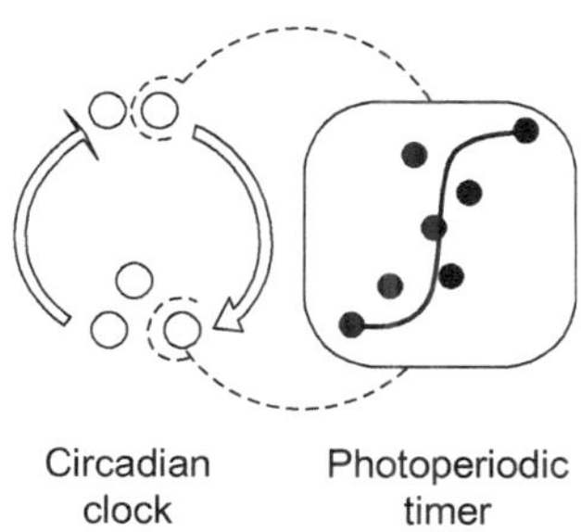

Fig. 5 The commensal model of biological timekeeping (commensal: one entity of the relationship benefits without cost or benefit to the other). In this reconciliation model, the photoperiodic timer co-opts one or a few of the output proteins of one or a few circadian clock genes, thus providing a time reference point to the timer. Communication with downstream photoperiodic time measurement would then occur without altering basic circadian clockworks. Downstream communication may be through the co-option of different titers of the same clock protein, or even different combinations of clock proteins in response to variable seasonal contexts. This model accommodates both stability of the circadian clock and independent evolution of timer and clock mechanisms.

protein, different proteins, or different combinations of proteins in response to selection imposed by seasonal contexts that vary in space and time.

In sum, the commensal model provides a genetic mechanism for the independent evolution of the daily circadian clock and the seasonal photoperiodic timer that incorporates evolution of photoperiodism in *Wyeomyia smithii* with more general conclusions and models across diverse insect taxa (Doležel, 2015; Goto, 2013; Koštál, 2011). Some gene or combination of genes in the circadian clock would then communicate with downstream processes, including photoperiodic time measurement, and those downstream processes can respond to selection without altering the basic circadian clockworks. The important remaining questions relate to (1) how alleles segregating in natural populations provide the genetic variation in and covariation between the circadian clock and the photoperiodic timer and (2) how this genetic architecture provides the opportunities for or impediments to their independent or joint evolution. The answer to these questions will continue to require not only molecular approaches but also approaches involving physiology, population genetics, and quantitative genetics, both within and among populations having evolved in the crucible of nature.

5. CONCLUSIONS

The evolutionary perspective provided by *Wyeomyia smithii* demonstrates that the genetic basis of photoperiodism and its relation to circadian rhythmicity is a population, not species, level consideration. The experiments and results discussed herein generate a model of the commensal relationship between the clock and the timer that permits canalization or homeostatic stability of the circadian clock and, at the same time, evolutionary flexibility of the photoperiodic timer.

The pioneering work of Danilevskii (1965) and colleagues treated photoperiodic response as a population-level character that, importantly, varied over geographic clines and with the degree of hybridity between populations. In spite of Danilevskii's broad vision, subsequent research into the mechanistic basis of photoperiodic response has generally emphasized detailed physiological and molecular studies of an investigator-specific species and population within that species. This research focus has generally led to the tacit assumptions that stage and formal properties of circadian–photoperiod relationships are a species-level character and that the specific population under consideration is representative of that entire species using but a single population (Saunders, 2010, Table 13.1). Although Pittendrigh did not challenge this narrow mind set, in his heart, he remained a true evolutionary biologist, interpreting variation in circadian rhythmicity, photoperiodism, and their connection "in the light of evolution" (Dobzhansky, 1964):

> *The last 25 years have yielded abundant proof that there is indeed a circadian component in the photoperiodic responses of a great many very different organisms; but we still lack any sure understanding of what that component is. We are not even sure it is always (indeed ever!) the clock that effects the photoperiodic time-measurement; in some species it may be the clock, and elsewhere not; and even when it does serve to measure photoperiod we cannot be sure it always does so in the same way. Moreover there is growing empirical support for our own intuitive preference that extensive convergent evolution probably underlies the phenomenological similarities between different taxa.*
>
> ***Pittendrigh, Elliott, and Takamura (1984, p. 37)***

The major contributions of research on photoperiodism in *Wyeomyia smithii* originate from this evolutionary perspective with the added dimension of focusing on multiple populations in their natural habitat: (1) to explore the genetic basis of photoperiodism in the context of historical biogeography through space and time by using multiple populations from over 100 localities in North America, (2) to intensify and clarify research on the circadian–photoperiodism connection by invoking the power of gene pleiotropy, and (3) to accelerate research on the genetics of photoperiodism by demonstrating the complexities of variation and covariation within and among populations of a single species. Among populations of *W. smithii* alone, these complexities encompass the range of variation associated with the entire pantheon of insect species. Even within a single population of *W. smithii*, there is in fact no "wild-type" photoperiodic response and every attribute of photoperiodic response or covariation with any putative underlying mechanism is critically dependent upon the genetic background in which it is assessed. A complex genetic architecture underlies photoperiodism within and among populations, underlies classic experiments to determine Pittendrigh's "circadian component," and therefore underlies a genetically and evolutionarily flexible interaction between photoperiodism and the circadian clock.

> *Evolutionary theory, which is our clearest guide to and explanation of unities in the living world also has another voice: It warns that functional similarity (functional unity) may often be the outcome of convergent evolution, of the fact that natural selection—Darwin's demon—is indifferent to precisely how the functional prerequisite is met.*
>
> ***Pittendrigh and Minis (1971, p. 245)***

Darwin's demon haunts those who advocate a genetic unity of photoperiodism or an invariant circadian–photoperiodic interaction, but delights those of us who pursue the diversity of natural systems and the intriguing variety of ways their underlying genetic architectures enable continuing adaption to the ever-changing natural world around them.

ACKNOWLEDGMENTS

Preparation of this manuscript was supported by NSF Grant DEB-1455506 (OPUS) to W.E.B. and C.M.H. based on our research over several decades of continual support from NSF programs in both DEB and IOS and from the Guggenheim and Fulbright Foundations. We are grateful for Zach Smith's collaboration in generating the data for Fig. 4.

REFERENCES

Andrewartha, H. G., & Birch, L. C. (1954). *The distribution and abundance of animals*. Chicago: The University of Chicago Press.

Armbruster, P. A., Bradshaw, W. E., & Holzapfel, C. M. (1998). Effects of postglacial range expansion on allozyme and quantitative genetic variation of the pitcher-plant mosquito, *Wyeomyia smithii*. *Evolution*, *52*, 1697–1704.

Beach, R. (1978). The required day number and timely induction of diapause in geographic strains of the mosquito *Aedes atropalpus*. *Journal of Insect Physiology*, *24*, 449–455.

Beavis, W. D. (1998). QTL analysis: Power, precision, and accuracy. In A. H. Patterson (Ed.), *Molecular dissection of complex traits* (pp. 145–162). Boca Raton, FL: CRC Press.

Bradshaw, W. E. (1976). Geography of photoperiodic response in a diapausing mosquito. *Nature*, *262*, 384–386.

Bradshaw, W. E. (1983). Interaction between the mosquito *Wyeomyia smithii*, the midge *Metriocnemus knabi*, and their carnivorous host *Sarracenia purpurea*. In J. H. Frank & L. P. Lounibos (Eds.), *Phytotelmata: Terrestrial plants as hosts of aquatic insect communities* (pp. 161–189). Medford, NJ: Plexus Publishing.

Bradshaw, W. E., & Creelman, R. A. (1984). Mutualism between the carnivorous purple pitcher plant and its inhabitants. *American Midland Naturalist*, *112*, 294–304.

Bradshaw, W. E., Emerson, K. J., Catchen, J. M., Cresko, W. A., & Holzapfel, C. M. (2012). Footprints in time: Comparative quantitative trait loci mapping of the pitcher-plant mosquito, *Wyeomyia smithii*. *Proceedings of the Royal Society B*, *279*, 4551–4558.

Bradshaw, W. E., Emerson, K. J., & Holzapfel, C. M. (2012). Genetic correlations and the evolution of photoperiodic time measurement within a local population of the pitcher-plant mosquito, *Wyeomyia smithii*. *Heredity*, *108*, 473–479.

Bradshaw, W. E., Haggerty, B. P., & Holzapfel, C. M. (2005). Epistasis underlying a fitness trait within a natural population of the pitcher-plant mosquito, *Wyeomyia smithii*. *Genetics*, *169*, 485–488.

Bradshaw, W. E., & Holzapfel, C. M. (1990). Evolution of phenology and demography in the pitcher-plant mosquito, *Wyeomyia smithii*. In F. Gilbert (Ed.), *Insect life cycles genetics, evolution and co-ordination* (pp. 47–67). London: Springer Verlag.

Bradshaw, W. E., & Holzapfel, C. M. (2001). Genetic shift in photoperiodic response correlated with global warming. *Proceedings of the National Academy of Sciences of the United States of America*, *98*, 14509–14511.

Bradshaw, W. E., & Holzapfel, C. M. (2006). Evolutionary response to rapid climate change. *Science*, *312*, 1477–1478.

Bradshaw, W. E., & Holzapfel, C. M. (2007). Evolution of animal photoperiodism. *Annual Review of Ecology, Evolution, and Systematics*, *38*, 1–25.
Bradshaw, W. E., & Holzapfel, C. M. (2008). Genetic response to rapid climate change: It's seasonal timing that matters. *Molecular Ecology*, *17*, 157–166.
Bradshaw, W. E., & Holzapfel, C. M. (2010a). What season is it anyway? Circadian tracking vs. photoperiodic anticipation in insects. *Journal of Biological Rhythms*, *25*, 155–165.
Bradshaw, W. E., & Holzapfel, C. M. (2010b). Light, time and the physiology of biotic response to rapid climate change in animals. *Annual Review of Physiology*, *72*, 147–166.
Bradshaw, W. E., & Holzapfel, C. M. (2010c). Insects at not so low temperature: Climate change in the temperate zone and its biotic consequences. In D. L. Denlinger & R. E. Lee Jr., (Eds.), *Low temperature biology of insects* (pp. 242–275). Cambridge: Cambridge University Press.
Bradshaw, W. E., Holzapfel, C. M., & Davison, T. E. (1998). Hourglass and rhythmic components of photoperiodic time measurement in the pitcher plant mosquito, *Wyeomyia smithii*. *Oecologia*, *117*, 486–495.
Bradshaw, W. E., Holzapfel, C. M., & Mathias, D. (2006). Circadian rhythmicity and photoperiodism in the pitcher-plant mosquito: Can the seasonal timer evolve independently of the circadian clock? *The American Naturalist*, *167*, 601–605.
Bradshaw, W. E., & Lounibos, L. P. (1972). Photoperiodic control of development in the pitcher-plant mosquito, *Wyeomyia smithii*. *Canadian Journal of Zoology*, *50*, 713–719.
Bradshaw, W. E., & Lounibos, L. P. (1977). Evolution of dormancy and its photoperiodic control in pitcher-plant mosquitoes. *Evolution*, *31*, 546–567.
Bradshaw, W. E., & Phillips, D. L. (1980). Photoperiodism and the photic environment of the pitcher-plant mosquito, *Wyeomyia smithii*. *Oecologia*, *44*, 311–316.
Bradshaw, W. E., Quebodeaux, M. C., & Holzapfel, C. M. (2003a). Circadian rhythmicity and photoperiodism in the pitcher-plant mosquito: Adaptive response to the photic environment or correlated response to the seasonal environment? *The American Naturalist*, *161*, 735–748.
Bradshaw, W. E., Quebodeaux, M. C., & Holzapfel, C. M. (2003b). The contribution of an hourglass timer to the evolution of photoperiodic response in the pitcher-plant mosquito, *Wyeomyia smithii*. *Evolution*, *57*, 2342–2349.
Bradshaw, W. E., Zani, P. A., & Holzapfel, C. M. (2004). Adaptation to temperate environments. *Evolution*, *58*, 1448–1462.
Buckley, H. L., Miller, T. E., Ellison, A. M., & Gotelli, N. J. (2003). Reverse latitudinal trends in species richness of pitcher-plant food webs. *Ecology Letters*, *6*, 825–829.
Buckley, H. L., Miller, T. E., Ellison, A. M., & Gotelli, N. J. (2010). Local to continental-scale variation in the richness and composition of an aquatic food web: Pitcher-plant food webs from local to continental scales. *Global Ecology and Biogeography*, *19*, 711–723.
Bünning, E. (1936). Die endonome Tagesrhythmik als Grundlage der photoperiodischen Reaktion. *Berichte der Deutschen Botanischen Gesellschaft*, *54*, 590–607.
Bünning, E. (1973). *The physiological clock*. New York: Springer Verlag.
Chevin, L.-M., & Lande, R. (2015). Evolution of environmental cues for phenotypic plasticity. *Evolution*, *69*, 2767–2775.
Danilevskii, A. S. (1965). *Photoperiodism and seasonal development in insects*. Edinburgh: Oliver and Boyd.
Danks, H. V. (2005). How similar are daily and seasonal biological clocks? *Journal of Insect Physiology*, *51*, 609–619.
Denlinger, D. L., Hahn, D. A., Merlin, C., Holzapfel, C. M., & Bradshaw, W. E. (2017). Keeping time without a spine: What can the insect clock teach us about seasonal adaptation? *Philosophical Transactions of the Royal Society of London. Series B, Biological sciences*, [in press].
De Vries, L. (1959). *German-English science dictionary*. New York: McGraw-Hill Book Company.

Dobzhansky, T. (1964). Biology, molecular and organismic. *American Zoologist*, *4*, 443–452.
Doležel, D. (2015). Photoperiodic time measurement in insects. *Current Opinion in Insect Science*, *7*, 98–103.
Emerson, K. J., Bradshaw, W. E., & Holzapfel, C. M. (2008). Concordance of the circadian clock with the environment is necessary to maximize fitness in natural populations. *Evolution*, *62*, 979–983.
Emerson, K. J., Bradshaw, W. E., & Holzapfel, C. M. (2009). Complications of complexity—Diapause in insects. *Trends in Genetics*, *25*, 217–225.
Emerson, K. J., Dake, S. J., Bradshaw, W. E., & Holzapfel, C. M. (2009). Evolution of photoperiodic time measurement is independent of the circadian clock in the pitcher-plant mosquito, *Wyeomyia smithii*. *Journal of Comparative Physiology A*, *195*, 385–391.
Emerson, K. J., Letaw, A. D., Bradshaw, W. E., & Holzapfel, C. M. (2008). Extrinsic light: dark cycles, rather than endogenous circadian cycles, affect the photoperiodic counter in the pitcher-plant mosquito, *Wyeomyia smithii*. *Journal of Comparative Physiology A*, *194*, 611–615.
Fisher, R. A. (1958). *The genetical theory of natural selection* (2nd ed.). . New York: Longman.
Franks, S. J., Sim, S., & Weis, A. E. (2007). Rapid evolution of flowering time by an annual plant in response to a climate fluctuation. *Proceedings of the National Academy of Sciences of the United States of America*, *104*, 1278–1282.
Fung, Y.-L. (1960). *A short history of Chinese philosophy*. New York: The Macmillan Company.
Gibson, G., & Dworkin, I. (2004). Uncovering cryptic genetic variation. *Nature Reviews Genetics*, *5*, 681–690.
Gomi, T., Nagasaka, M., Fukuda, T., & Hagihara, H. (2007). Shifting of the life cycle and life-history traits of the fall webworm in relation to climate change. *Entomologia Experimentalis et Applicata*, *125*, 179–184.
Gorham, E., Lehman, C., Dyke, A., Janssens, J., & Dyke, L. (2007). Temporal and spatial aspects of peatland initiation following deglaciation in North America. *Quaternary Science Reviews*, *26*, 300–311.
Goto, S. G. (2013). Role of circadian clock genes in insect photoperiodism. *Entomological Science*, *16*, 1–16.
Halsey, L. A., Vitt, D. H., & Gignac, L. (2000). *Sphagnum*-dominated peatlands in North America since the last glacial maximum: Their occurrence and extent. *Bryologist*, *103*, 334–352.
Hard, J. J., Bradshaw, W. E., & Holzapfel, C. M. (1992). Epistasis and the genetic divergence of photoperiodism between populations of the pitcher-plant mosquito, *Wyeomyia smithii*. *Genetics*, *131*, 389–396.
Hard, J. J., Bradshaw, W. E., & Holzapfel, C. M. (1993). The genetic basis of photoperiodism and evolutionary divergence among populations of the pitcher-plant mosquito, *Wyeomyia smithii*. *The American Naturalist*, *142*, 457–473.
Hayman, B. L. (1958). The separation of epistatic from additive and dominance variation in generation means. *Heredity*, *12*, 371–390.
Hayman, B. L. (1960a). The separation of epistatic from additive and dominance variation in generation means. II. *Genetics*, *31*, 133–146.
Hayman, B. L. (1960b). Maximum likelihood estimate of genetic components of variation. *Biometrics*, *16*, 369–381.
Hughes, L. (2000). Biological consequences of global warming: Is the signal already apparent? *Trends in Ecology & Evolution*, *15*, 56–61.
Hut, R. A., Paolucci, S., Dor, R., Kyriacou, C. P., & Daan, S. (2013). Latitudinal clines: An evolutionary view on biological rhythms. *Proceedings of the Royal Society B*, *280*. 20130433.
IPCC. (2007). *Climate change 2007: The physical basis. Contribution of working group I to the fourth assessment of the intergovernmental panel on climate change*. Geneva: IPCC Secretariat.

Istock, C. A., & Weisburg, W. G. (1987). Strong habitat selection and the development of population structure in a mosquito. *Evolutionary Ecology*, *1*, 348–362.

Karl, T. R., & Trenberth, K. E. (2003). Modern global climate change. *Science*, *302*, 1719–1723.

Koŝtál, V. (2011). Insect photoperiodic calendar and circadian clock: Independence, cooperation, or unity? *Journal of Insect Physiology*, *57*, 538–556.

Kubota, A., Kita, S., Ishizaki, K., Nishihama, R., Yamato, K. T., & Kohchi, T. (2014). Co-option of a photoperiodic growth-phase transition system during land plant evolution. *Nature Communications*, *5*, 3668.

Lair, K. P., Bradshaw, W. E., & Holzapfel, C. M. (1997). Evolutionary divergence of the genetic architecture underlying photoperiodism in the pitcher-plant mosquito, *Wyeomyia smithii*. *Genetics*, *147*, 1873–1883.

Lankinen, P., & Forsman, P. (2006). Independence of genetic geographical variation between photoperiodic diapause, circadian eclosion rhythm, and Thr-Gly repeat region of the *period* gene in *Drosophila littoralis*. *Journal of Biological Rhythms*, *21*, 3–12.

Lucas-Reina, E., Romero, F. J., Romero, J. M., & Valverde, F. (2015). An evolutionarily conserved DOF-CONSTANS module controls plant photoperiodic signaling. *Plant Physiology*, *168*, 561–574.

Lynch, M., & Walsh, B. (1998). *Genetics and the analysis of quantitative traits*. Sunderland, MA: Sinauer Associates.

MacArthur, R. H. (1972). *Geographical ecology*. New York: Harper & Row.

Mackay, T. F. C., Stone, E. A., & Ayroles, J. F. (2009). The genetics of quantitative traits: Challenges and prospects. *Nature Reviews Genetics*, *10*, 565–577.

Mather, K., & Jinks, J. L. (1982). *Biometrical genetics*. London: Chapman and Hall.

Mathias, D., Jacky, L., Bradshaw, W. E., & Holzapfel, C. M. (2007). Quantitative trait loci associated with photoperiodic response and stage of diapause in the pitcher-plant mosquito, *Wyeomyia smithii*. *Genetics*, *176*, 391–402.

Mayr, E. (1954). Change of genetic environment and evolution. In J. Huxley, A. C. Hardy, & E. B. Ford (Eds.), *Evolution as a process* (pp. 157–180). London: Allen & Unwin.

Merz, C., Catchen, J. M., Hanson-Smith, V., Emerson, K. J., Bradshaw, W. E., & Holzapfel, C. M. (2013). Replicate phylogenies and post-glacial range expansion of the pitcher-plant mosquito, *Wyeomyia smithii*, in North America. *PLoS One*, *8*. e72262.

Meuti, M. E., & Denlinger, D. L. (2013). Evolutionary links between circadian clocks and photoperiodic diapause in insects. *Integrative and Comparative Biology*, *53*, 131–143.

Moeur, J. E., & Istock, C. A. (1982). Chromosomal polymorphisms in the pitcher-plant mosquito, *Wyeomyia smithii*. *Chromosoma*, *84*, 623–651.

Nelson, R. J., Denlinger, D. L., & Somers, D. E. (2010). *Photoperiodism*. New York: Oxford University Press.

Numata, H., Mihazaki, Y., & Ikeno, T. (2015). Common features in diverse insect clocks. *Zoological Letters*, *1*, 10.

Omura, S., Numata, H., & Goto, S. G. (2016). Circadian clock regulates photoperiodic responses governed by distinct output pathways in the bean bug, *Riptortus pedestris*. *Biological Rhythm Research*, *47*, 937–945.

Paaby, A. B., & Gibson, G. (2016). Cryptic genetic variation in evolutionary developmental genetics. *Biology*, *5*, 28.

Paaby, A. B., & Rockman, M. V. (2014). Cryptic genetic variation: Evolution's hidden substrate. *Nature Reviews Genetics*, *15*, 247–258.

Parmesan, C., & Yohe, G. (2003). A globally coherent fingerprint of climate change impacts across natural systems. *Nature*, *421*, 37–42.

Peñuelas, J., & Filella, I. (2001). Response to a warming world. *Science*, *294*, 793–795.

Phillips, P. C. (2005). Testing hypotheses regarding the genetics of adaptation. *Genetica*, *123*, 15–24.

Phillips, P. C. (2008). Epistasis—The essential role of gene interactions in the structure and evolution of genetic systems. *Nature Reviews Genetics*, *9*, 855–867.
Pittendrigh, C. S. (1966). The circadian oscillation in *Drosophila pseudoobscura* pupae: A model for the photoperiodic clock. *Zeitschrift für Pflanzenphysiologie*, *54*, 275–307.
Pittendrigh, C. S. (1972). Circadian surfaces and the diversity of possible roles of circadian organization in photoperiodic induction. *Proceedings of the National Academy of Sciences of the United States of America*, *69*, 2734–2737.
Pittendrigh, C. S. (1981). Circadian organization and the photoperiodic phenomena. In B. K. Follet & D. E. Follett (Eds.), *Biological clocks in seasonal reproductive cycles* (pp. 1–35). Bristol: Wright.
Pittendrigh, C. S., Elliott, J., & Takamura, T. (1984). The circadian component in photoperiodic induction. In R. Porter & J. M. Collins (Eds.), *Photoperiodic regulation of insect and molluscan hormones* (pp. 26–47). London: Pitman.
Pittendrigh, C. S., & Minis, D. H. (1964). The entrainment of circadian oscillations by light and their role as photoperiodic clocks. *The American Naturalist*, *98*, 261–294.
Pittendrigh, C. S., & Minis, D. H. (1971). The photoperiodic time measurement in *Pectinophora gossypiella* and its relation to the circadian system in that species. In M. Menaker (Ed.), *Biochronometry* (pp. 212–250). Washington, DC: National Academy of Sciences.
Roff, D. A. (1997). *Evolutionary quantitative genetics*. New York: Chapman & Hall.
Root, T. L., Price, J. T., Hall, K. R., Schneider, S. H., Rosenzweig, C., & Pounds, J. A. (2003). Fingerprints of global warming on wild animals and plants. *Nature*, *421*, 57–60.
Saunders, D. S. (1981). Insect photoperiodism—The clock and the counter: A review. *Physiological Entomology*, *6*, 99–116.
Saunders, D. S. (2002). Circadian rhythms in photoperiodism. In C. G. H. Steel, X. Vafopoulou, & R. D. Lewis (Eds.), *Insect clocks* (3rd ed., pp. 339–375). Amsterdam: Elsevier Science.
Saunders, D. S. (2009). Circadian rhythms and the evolution of photoperiodic timing in insects. *Physiological Entomology*, *34*, 301–308.
Saunders, D. S. (2010). Photoperiodism in insects: Migration and diapause responses. In R. J. Nelson, D. L. Denlinger, & D. E. Somers (Eds.), *Photoperiodism: The biological calendar* (pp. 218–257). New York: Oxford University Press.
Saunders, D. S. (2013). Insect photoperiodism: Measuring the night. *Journal of Insect Physiology*, *59*, 1–10.
Saunders, D. S. (2016). The temporal 'structure' and function of the insect photoperiodic clock: A tribute to Colin S. Pittendrigh. *Physiological Entomology*, *41*, 1–18.
Saunders, D. S., & Bertossa, R. C. (2011). Deciphering time measurement: The role of circadian 'clock' genes and formal experimentation in insect photoperiodism. *Journal of Insect Physiology*, *57*, 557–566.
Smith, J. B. (1902). Life history of Aëdes smithii *Coq. Journal of the New York Entomological Society*, *10*, 10–15.
Stott, P. A., Tett, S. F. B., Jones, G. S., Allen, M. R., Mitchell, J. F. B., & Jenkins, G. J. (2000). External control of 20th century temperature by natural and anthropogenic forcings. *Science*, *290*, 2133–2137.
Taylor, F. (1980). Optimal switching to diapause in relation to the onset of winter. *Theoretical Population Biology*, *18*, 125–133.
Taylor, F. (1985). Estimating the ends of the sensitive period for diapause induction in arthropods. *Journal of Theoretical Biology*, *117*, 319–336.
Tormey, D., Colbourne, J. K., Mockaitis, K., Choi, J.-H., Lopez, J., Burkhart, J., et al. (2015). Evolutionary divergence of core and post-translational circadian clock genes in the pitcher-plant mosquito, *Wyeomyia smithii*. *BMC Genomics*, *16*, 754.

Urbansky, J., Mogi, M., O'Donnell, D., DeCotiis, M., Toma, T., & Armbruster, P. (2012). Rapid adaptive evolution of photoperiodic response during invasion and range expansion across a climatic gradient. *The American Naturalist*, *179*, 490–500. Supplemental data from Japanese populations. http://datadryad.org/resource/doi:10.5061/dryad.68277435.
Wallace, B. (1991). Coadaptation revisited. *Journal of Heredity*, *82*, 89–95.
Wegis, M. C., Bradshaw, W. E., Davison, T. E., & Holzapfel, C. M. (1987). Rhythmic components of photoperiodic time measurement in the pitcher-plant mosquito, *Wyeomyia smithii*. *Oecologia*, *110*, 32–39.
Wright, S. (1968). Evolution and the genetics of populations. *Genetic and biometric foundations: Vol. 1*. Chicago: University of Chicago Press.
Wright, S. (1977). Evolution and the genetics of populations. *Experimental results and evolutionary deductions: Vol. 3*. Chicago: University of Chicago Press.
Xu, S. (2003). Theoretical basis of the Baevis effect. *Genetics*, *166*, 2259–2268.
Yadav, A., Dhole, K., & Sinha, H. (2016). Differential regulation of cryptic genetic variation shapes the interactome underlying complex traits. *Genome Biology and Evolution*, *8*, 3559–3573.
Zani, P. A., Cohnstaedt, L. W., Corbin, D., Bradshaw, W. E., & Holzapfel, C. M. (2005). Reproductive value in a complex life cycle: Heat tolerance of the pitcher-plant mosquito, *Wyeomyia smithii*. *Journal of Evolutionary Biology*, *18*, 101–105.

CHAPTER THREE

Natural Variation in Human Clocks

Malcolm von Schantz[1]
Faculty of Health and Medical Sciences, University of Surrey, Guildford, Surrey, United Kingdom
[1]Corresponding author: e-mail address: m.von.schantz@surrey.ac.uk

Contents

Abstract

Our own species has a diurnal activity pattern and an average circadian period of 24.2 h. Exact determination of circadian period requires expensive and intrusive protocols, and investigators are therefore using chronotype questionnaires as a proxy quantitative measure. Both measures show a normal distribution suggestive of a polygenic trait. The genetic components of the 24-h feedback loop that generates circadian rhythms within our cells have been mapped in detail, identifying a number of candidate genes which have been investigated for genetic polymorphisms relating to the phenotypic variance. Key in this mechanism is the inhibitory complex containing period and cryptochrome proteins and interacting protein kinases and ubiquitin ligases, and the stability of this complex is recognized as the major determinant of circadian periodicity. The identification of the causative mutations in familial circadian rhythms sleep disorders has shed additional light into this mechanism. Mutations in the negative feedback protein-encoding genes *PER2* and *CRY2* as well as the *CSNK1D* gene encoding casein kinase I delta have been shown to cause advanced sleep phase disorder, and a mutation in the *CRY1* gene delayed sleep phase disorder. The candidate gene approach has also yielded a number of genetic associations with chronotype as determined by questionnaires. More recently, genome-wide association studies of chronotype have both confirmed associations with the candidate clock gene *PER2* and identified a serious of novel genes associated with variability in circadian rhythmicity, which have yet to be explored.

Advances in Genetics, Volume 99
ISSN 0065-2660
https://doi.org/10.1016/bs.adgen.2017.09.003

While considerable progress has thus been made with mapping the phenotypic diversity in human circadian rhythms and the genomic variability that causes it, studies to date have been mostly focused on individuals of European descent, and there is a strong need for research on other populations.

1. INTRODUCTION

The transition between day and night on our planet is defined by the presence or absence of solar illumination. A multitude of other parameters are dependent on this, such as temperature, evaporation rate, relative humidity, etc. (Park, 1940). Secondarily, the respective advantages of the diurnal and nocturnal niches are affected by the prevalence of predators and/or prey. These differences can be dramatic, and it is not surprising, therefore, that most animals, both vertebrate and invertebrate, have adapted their physiology and behavior specifically for either diurnality or nocturnality. Such adaptations and readaptations have occurred several times in evolution within relatively short time spans. The most famous one is the so-called nocturnal bottleneck, through which mammals emerged from their diurnal reptilian ancestors some 200 million years ago (Smale, Lee, & Nunez, 2003). This transition enabled them to exploit an ecological niche that was virtually unused by animals of the same size, and was accompanied by a number of radical changes, including in their metabolism (Crompton, Taylor, & Jagger, 1978), visual (von Schantz, 1998), and circadian systems (Menaker, Moreira, & Tosini, 1997). A minority of mammals in a number of lineages (Park, 1940), including our own, have since readapted to a diurnal lifestyle.

These adaptations are not cultural, but genetically programmed. Prominently, they include specific features in the circadian system. The classical Aschoff's rule (Aschoff, 1958, 1960, 1979) states that, in constant darkness, free-running period length (τ) for nocturnal species is shorter than 24 h, and for diurnal species, longer than 24 h (in addition to a number of other predictions about the animals' behavior in different light regimes). Although particularly diurnal species are subject to variability and exceptions (Smale et al., 2003), this rule remains generally valid for mammals.

It is well established that our own species originated in Africa. Until recently, the oldest known exemplar fossils of anatomically modern *Homo sapiens*, dating back 200,000 years, were the ones found in the Omo valley of Ethiopia. However, more recently described findings from Jebel Irhoud, Morocco, have moved the age of the earliest known human fossils back to 315,000 years ago (Hublin et al., 2017; Richter et al., 2017), and created greater uncertainty about the specific area in which our species first appeared. We are certain that this was within the African continent, but we are less certain than we thought we were as to whether this took place closer to the equator or closer to the Topic of Cancer. With certainty, however, no members of our species encountered extreme photoperiods until after part of the human population emigrated to Eurasia across the Arabian peninsula. Here, too, the recent consensus that successful emigration occurred during a limited time window around 70,000 years ago (Tucci & Akey, 2016) has recently been challenged by suggestions of traces of a much earlier emigration some 270,000 years ago (Posth et al., 2017).

The humans who had migrated to Eurasia adapted to new environmental factors, partially by gaining genetic material through interbreeding with two related species of archaic humans that had already been living there much longer, Neanderthals and Denisovans (Rogers Ackermann, Mackay, & Arnold, 2015). These included the gradual loss of skin pigmentation in populations which established themselves at Northern latitudes (Elias & Williams, 2015). The last landmass to be colonized was the Americas. Here, humans arrived 14–15,000 years ago (Jenkins et al., 2012) from Siberia via Bering's strait. The subsequent colonization of the landmass in a southerly direction toward the equator and beyond was thus relatively rapid. As a result of this gradual conquest of the continents, our species now inhabits the entire planet, with permanent settlements ranging from Alert on Ellesmere Island, Canada (82°28′ N) to Puerto Toro on Navarino Island, Chile (55°05′ S). Thus, the photoperiods experienced by our species range from no variation at the equator to a cycling between midnight sun and polar night interspersed by 1 month of twilight in spring and one in autumn in Alert.

2. DIRECT MEASUREMENT OF HUMAN PERIOD

Circadian period in other animals is almost invariable determined by monitoring free-running activity rhythms. Scientists working with human subjects have the luxury of being able to collect direct physiological measures (core body temperature and/or melatonin) and even repeated blood samples, and are able to request their subjects to remain immobile during constant routine protocols, but also face the complication that studying their experimental organism under constant conditions is very costly both in terms of remuneration to the volunteers and in terms of paying the cost of the investigation unit and its staff.

The first published study of free-running human period length (Aschoff & Wever, 1962) was performed in the famous bunker in Andechs. The reported periodicity of 25 h, as we now know, was artificially lengthened by the subject being able to control their own light–dark cycle. It took until the end of the century for a more reliable estimate, based on a larger number of individuals (24) to be published (Czeisler et al., 1999). Rather than bringing the subjects into constant darkness, which would have solved the problem experienced at Andechs but made the study protocol very difficult to tolerate for a period of nearly a month, the Harvard group applied the forced desynchrony protocol (Kleitman, 1939), in which subjects are exposed to a photoperiod with a day length that is either too long or too short to allow entrainment, causing a desynchronization of circadian rhythms from the subject's sleep–wake cycle. Using this paradigm, average human period length was determined to be 24.18 h, with an observed range between 23.52 and 24.29 h in 24 subjects. Two age groups were included in the study (mean ages 23.7 and 67.4 years, respectively). No significant differences were observed between these groups, and it was concluded that, as observed in other species, human circadian period does not change appreciably with age. Our own group subsequently performed a period measurement using a similar paradigm, obtaining a very similar average period measure (24.16 ± 0.17 h) (Hasan et al., 2012).

A potential alternative to the disruption of placing human volunteers in a clinical investigation unit for several weeks is to study blind individuals who lack functioning photoreceptors and whose circadian rhythms are constantly free-running as they lack the ability to entrain in response to the external light:dark cycle. Thus, precise measures of free-running circadian period can be made in these individuals by analyzing either the timing of melatonin secretion (via saliva samples) or its secretion product 6-sulfatoxymelatonin in urine. We reported data from 26 such individuals (Robilliard et al., 2002), with a range between 23.92 and 24.95 h and an average of 24.53 ± 0.25 h. The longer periods observed with this method than with forced desynchrony may reflect either an artifact of either method, and/or a selection bias in the free-running individuals (a free-running individual with a near-24-h period may not be distinguishable from an entrained one).

3. PROXIES OF HUMAN PERIOD

It will be clear from the previous section that direct measurement of human periodicity is costly and intrusive, and not feasible at a larger scale;

certainly not at the scale needed for genetic analysis. Therefore, investigators have developed proxies for circadian period, variously known as diurnal preference or chronotype (see Adan et al., 2012, for a comprehensive review). The measure that is best understood in biological terms is the Morningness–Eveningness questionnaire (MEQ) developed more than 40 years ago (Horne & Östberg, 1976). The questionnaire consists of 19 multiple choice items, generally asking about preferred timing of activities such as rise and bed times. Each item has four options with different weighting, resulting in a total score range between 16 (highest eveningness) and 86 (highest morningness). Distribution of score in most population samples is close to normal. The original publication proposed a typology with specific score ranging defining different classes of morning and evening types. This typology has aged much less well than the questionnaire itself. The participants in the original study were young (18–32), while there is an advance in circadian phase, wake time, and MEQ score with increasing age (Duffy & Czeisler, 2002). It has also been reported that, while in rural communities with a more traditional lifestyle characterized by generally earlier bed and rise times, the distribution pattern and age-dependency of MEQ are similar, but shifted significantly toward higher scores representing greater morningness (von Schantz et al., 2015). Thus, a moderate morning type living in a metropolitan area could have the same MEQ score as a moderate evening type in a rural setting. The variability with age and locality complicates the use of MEQ score in genetic analyses.

On the other hand, MEQ score has been validated to correlate with circadian period (Duffy, Rimmer, & Czeisler, 2001). However, in addition to this circadian component, MEQ has been found to harbor a component of sleep homeostasis, with morning vs evening preference, resulting from differences in the build-up or dissipation of homeostatic sleep pressure (Mongrain, Carrier, & Dumont, 2006; Taillard, Philip, Coste, Sagaspe, & Bioulac, 2003). It appears possible that the proportion between circadian and sleep homeostatic influence over MEQ score varies between individuals. Although, as noted earlier (Czeisler et al., 1999), human circadian period is not thought to change across the life span, there is an age-dependence in chronotype resulting in gradually increasing morningness (Duffy et al., 2001), with an increase in MEQ score by 1 unit approximately every 3.8 years (Robilliard et al., 2002; von Schantz et al., 2015). This phenomenon, which is also observed with other measures of chronotype (Roenneberg, Wirz-Justice, & Merrow, 2003) is likely attributable to the sleep homeostatic component.

The main difference between MEQ and the other frequently used measure of chronotype, the Munich Chronotype Questionnaire (MCTQ) (Roenneberg et al., 2003), is that while the former asks about preferred timing of specific activities, the latter asks about actual habitual timings, distinguishing between workdays and weekends. Therefore, the MCTQ has the clear advantage that its core only consists of 12 short questions, all of which are answered with a simple clock time. Because responding to it takes much less time than the MEQ, it is more amenable to large data collections, including using online forms.

The two questionnaires most in use, the MEQ and the MCTQ complement and have been validated (Zavada, Gordijn, Beersma, Daan, & Roenneberg, 2005) against each other, the former being more trait-like and the latter being more state-like and allowing the calculation of habitual sleep duration and so-called social jetlag, e.g., the difference in sleep phase between weekdays and free days (Wittmann, Dinich, Merrow, & Roenneberg, 2006). It is logical, therefore, that studies of the heritability of chronotype have focused on the MEQ, with a range between 21% (Evans et al., 2011) and 38% (von Schantz et al., 2015) observed in family studies, and 52% in a twin study (Barclay, Eley, Buysse, Archer, & Gregory, 2010). This suggests that this is a quantitative trait that is amenable for genetic analysis, as will be discussed later.

4. EXTREMES OF HUMAN CIRCADIAN PERIOD

The diagnoses advanced and delayed sleep phase disorder (ASPD and DSPD) are defined as circadian rhythm sleep disorder and are used to describe conditions where patients are unable to conform to the societal norms of sleep and wake-up times, and have an intrinsic preference for timings that are either significantly earlier or significantly later (by 3 h or more) (American Association of Sleep Medicine, 2014; Zee, Attarian, & Videnovic, 2013). These conditions could be described as extremely early and extremely late chronotypes, and potentially be explained in terms of extremely short or extremely long circadian periods. Initially, at least, ASPD appeared more amenable to such an explanation. It has also been found in a number of instances to be inherited in a Mendelian pattern, and large pedigrees of families with patients have been described (Jones et al., 1999). A patient from one of these was studied under free-running conditions for a period of 18 days, and found to display a period in sleep–wake cycle and core body temperature of 23.3 h. Very discrepant figures have been reported for the incidence of DSPD, ranging between 0.2% and 10% (Zee et al., 2013). This partially reflects heterogeneity in diagnostic criteria (Micic et al., 2016), and partially the transient nature of the condition in many individuals. Humans show a sharp peak in eveningness which occurs in adolescence around 20 years of age (Roenneberg et al., 2004), which is followed by a gradual and steady decrease. The wide difference in the diagnostic criteria actually applied has been viewed as consistent with a more heterogeneous etiology. Earlier reports suggested more complex patterns of inheritance (Ancoli-Israel, Schnierow, Kelsoe, & Fink, 2001) and firm evidence for Mendelian inheritance only emerged earlier this year (Patke et al., 2017). One measurement of period length in sleep–wake cycle and core body temperature from a patient kept in temporal isolation for 17 days has been reported—25.38 h (Campbell & Murphy, 2007). More recently, a period length of 24.8 h and a diminished amplitude in core body temperature were reported in a DSPD patient (Patke et al., 2017). Thus, both ASPD and DSPD have a recognizable phenotype compatible with an extreme period and chronotype. There is also a clear possibility that cases of both may represent disorders of entrainment, either entirely or in combination with an altered circadian period. The fact that, at least in some instances, they are inherited in a monogenic pattern makes them promising candidates for genetic analysis—a promise that has been delivered to some considerable extent.

5. GENETIC CONTROL OF CIRCADIAN RHYTHMS IN MAMMALS

Although we now know that circadian rhythms are also present in red blood cells, which lack nuclei (O'Neill & Reddy, 2011), it is generally accepted that a group of specific clock genes are critically responsible for the generation and control of most circadian processes in humans and other mammals. We also know from an elegant set of experiment that behavioral circadian period in mammals is determined by genes expressed in the suprachiasmatic nuclei (SCN) of the hypothalamus (Welsh, Takahashi, & Kay, 2010). Crosswise transplantations of SCNs from wild-type hamsters and hamsters from the *tau* mutant, which has a supershort free-running circadian period (20 h in homozygotes), conveyed a rest–activity phenotype dictated by the genotype of the SCN, regardless of the genotype of the rest of the body (Ralph, Foster, Davis, & Menaker, 1990). The SCN is strategically located just above the optic chiasm, from which it receives retinal input informing it of the external light–dark cycle, and exerts its outputs through neuronal firing and the secretion of humoral factors (Welsh et al., 2010).

Through forward and reverse genetics approaches, the key genetic components of the mammalian clock have been established (for recent reviews on genomic, transcriptional, and posttranscriptional mechanisms, see Papazyan, Zhang, & Lazar, 2016; Takahashi, 2017; and Preussner & Heyd, 2016, respectively). A key to the rapid breakthrough in mammalian clock genetics during the final years of the last century was the remarkable realization that most of the major classes of molecular components in the mammalian clock are orthologous with those of the *Drosophila* clock, which was already quite well understood, the major difference being that, following two genome duplications in the early vertebrate lineage (Ohno, 1970), mammals have two or three paralogs of key components whereas *Drosophila* only has one (Looby & Loudon, 2005; Tauber, Last, Olive, & Kyriacou, 2004; von Schantz, Jenkins, & Archer, 2006). It is quite remarkable how well the model of interacting positive and negative regulators at the core of the mammalian clock machinery proposed two decades ago has lasted, apart from the early mistake in misidentifying a mammalian homolog of the *Drosophila timeless* gene as a key negative regulator (Gotter et al., 2000).

At the center of the mammalian clock gene machinery is a heterodimer of two transcription factor proteins, most commonly CLOCK and BMAL1 (the product of the *Arntl1 gene*). CLOCK may be interchanged with its paralog NPAS2 [particularly in the forebrain (Reick, Garcia, Dudley, & McKnight, 2001)], while BMAL2 (encoded by the *Arntl2* gene) may be interchangeable with BMAL1 (Hogenesch et al., 2000; Shi et al., 2010). The proteins forming this dimer contain basic helix–loop–helix domains, with basic amino acids binding to the acidic residues of DNA, typically at a six-base consensus motif called *E*-box (for enhancer box). Critical for the generation of circadian rhythms is the binding of this dimer to E-boxes in the promoter regions of the genes of the *Cryptochrome* (*Cry*) and *Period* (*Per*) families, specifically *Cry1* and *Cry2* and *Per1* and *Per2* (there is also a *Per3* gene, although the evidence suggests that it is nonessential for circadian clock function Bae et al., 2001 and more important for sleep homeostasis Viola et al., 2007; see further in Section 8). The genes are widely expressed across the body, but with overall control by signals from the SCN. Starting in the early morning, transcripts from these promoters start to accumulate in the cells of the SCN, followed by translation into CRY and PER proteins that accumulate over the afternoon and with a peak in the evening. This is followed by the repression phase, during which these proteins interact with the serine/threonine kinases casein kinase 1∂ and 1ε and glycogen synthase kinase 3 (GSK3) in a manner that ensures their coordinated import into the cell nucleus (Gallego & Virshup, 2007). Here, they form a large, 1-MDa multiprotein complex together with other proteins (Brown et al., 2005), which exerts negative feedback inhibition of the transactivator activity of the Clock–BMAL1 complex. This leads to the cessation of synthesis of *Cry* and *Per* RNA, and, consequentially, CRY and PER protein. The repression phase is ended and the loop closed by the removal of the inhibitory CRY–PER complex. Phosphorylation of the PER proteins by casein kinase I recruits F-box ubiquitin ligase proteins that mediate their proteasome-mediated degradation (Eide et al., 2005; Shirogane, Jin, Ang, & Harper, 2005). With levels of *Cry* and *Per* RNA and protein in the cell back to zero, a new cycle can begin roughly 24 h later.

Studies in a number with a number of different mutants consistently indicate that circadian period length is regulated by the stability of the repressor (Takahashi, 2017). This, in turn, is determined by the ability of the PER and CRY proteins to interact in various specific ways with the kinases and ubiquitin ligases. This interaction determines how quickly the clock proteins will be destroyed. The *tau* hamster phenotype with the supershort circadian period mentioned earlier, for example, is caused by a mutation in casein kinase Iε. A mutated kinase could be assumed to be a loss-of-function mutant that would decrease phosphorylation of the substrate protein, thereby increasing its lifespan and thus the circadian period length. However, in this instance, this is a gain-of-function mutation insofar as it actually increases the ability of the kinase to phosphorylate PER proteins in positions that enhance their interaction with ubiquitin ligases and thus speed up their destruction (Gallego, Eide, Woolf, Virshup, & Forger, 2006).

Additional to this central loop of expression is an interlocked one, though which the CLOCK–BMAL1 complex also actives the expression of the nuclear receptors REV-ERBα and REV-ERBβ (enclosed by the genes *Nr1d1* and *Nr1d2*). These proteins, in turn, repress the expression of BMAL1 and CLOCK (Liu et al., 2008; Preitner et al., 2002) through binding to ROR DNA elements.

The functions of the main clock genes have been probed by the creation of knockout mice. The only single knockout that abolished circadian rhythms under free-running conditions has been *Bmal1* (Bunger et al., 2000). Single knockouts of other core clock single clock genes have generally led to altered periodicity and stability of free-running circadian rhythms. Double knockouts of *Per1* and *Per2* (Bae et al., 2001) and *Cry1* and *Cry2* (van der Horst et al., 1999) abolished free-running rhythmicity, however, similarly to the *Bmal1* knockout. As will be seen in the next section, gain-of-function point mutations in key clock components can have even more dramatic consequences than knocking out a particular gene.

6. GENETIC STUDIES OF CIRCADIAN RHYTHM SLEEP DISORDERS

Given that they represent extremes of phenotype, and that they show Mendelian inheritance in a number of pedigrees, it is not entirely surprising that the search for genetic associations of ASPD and DSPD has been such a success story. It is perhaps more remarkable that these successes have all been achieved using the candidate gene approach with known clock genes, making this quite possibly the area where this approach has been most fruitful in human behavioral genetics.

Although it is not essential for circadian rhythmicity, a number of polymorphisms in the *PERIOD3* (*PER3*) gene have been reported as protective and risk factors for simplex cases of DSPD including a haplotype defined by a missense polymorphism (rs10462020) (Ebisawa et al., 2001), a variable number tandem repeat (VNTR) polymorphism in the coding region (rs57875989) (Archer et al., 2003), and a complex promoter polymorphism haplotype (Archer et al., 2010). A *PER3* haplotype has also been suggested to cause ASPD with depression in a human pedigree of four individuals (Zhang et al., 2016).

More remarkable success, however, has been achieved with a number of pedigrees with familial ASPD and, recently, with one DSPD pedigree, where the conditions are inherited in a Mendelian, autosomal dominant fashion. The first such report described the causative mutation the large ASPD pedigree previously reported from Utah, where the period of one patient had been measured as 23.3 h (Jones et al., 1999). This pedigree was large enough to permit classical linkage analysis, pointing toward the telomeric region of the short arm of chromosome 2, which contains the human *PER2* gene (Toh et al., 2001). Exons were screened using single-strand conformation polymorphism analysis, pinpointing exon 17 as the site of the mutation. Here, a missense mutation (rs121908635) was identified which cosegregated with the trait. This mutation was within the predicted site of interaction with casein kinase I, replacing a serine residue with a glycine. In vitro expression of the mutant allele confirmed that it was hypophosphorylated compared to wild type. Further confirmation was provided by the generation of a knock-in mouse model (Xu et al., 2007). Bred into a *Per2* knockout background, mice carrying the humanized mutated gene showed a period length of 20.7 h and an activity onset 4 h earlier than wild type.

A second ASPD pedigree was investigated by the same team (Xu et al., 2005). This pedigree had a very similar phenotype as the one carrying the *PER2* mutation. Because it was much smaller, linkage analysis was not an option, and mutation screening was therefore performed in candidate clock genes using denaturing HLPC and direct sequencing. Using this strategy, a mutation (rs104894561) was discovered in the *CSNK1D* gene encoding casein kinase I∂, which caused a similarly shortened period in knock-in mice. This is believed to be a gain-of-function mutation similar to the mutation in the casein kinase 1ε paralog in the *tau* hamster as described earlier (Gallego et al., 2006).

The latest familial ASPD mutation reported by this group (Hirano et al., 2016) is based on an even smaller pedigree with a mere three affected individuals exhibiting a strong phenotype where dim light melatonin onset was 16:41 h (compared to a population mean of 20:50 h Burgess & Fogg, 2008). Here, yet again, the approach of screening candidate genes was chosen—no less than 26 of these were analyzed. In this instance, a missense polymorphism (rs201220841) was discovered in the *CRY2* gene that was found to affect the ability of the CRY2 protein to interact with the E3 ubiquitin ligase FBXL3. The period of knock-in mice carrying the mutation was significantly shorter, although the difference was surprisingly small (23:52 vs 23:70 for knock-in mice carrying the wild type human *CRY2* gene and 23:74 for wild type mice). Using luciferase reporter transgenes, shortened periodicity was also observed in peripheral oscillators. Mice carrying the mutation also showed attenuated phase delays in response to light pulses that were less than half of either wild type, while light-pulse-induced phase advances were of normal length. Thus, is appears that in this instance, the condition may be caused by the combination of a shorter than average circadian period and a reduced efficiency in compensating for this by phase delay.

Until earlier this year, only polymorphisms conveying risk for or protection against spontaneous DSPD had been reported. This imbalance has now been rectified by a report on a Turkish family where the condition is inherited in a Mendelian fashion (Patke et al., 2017). The phenotype was characterized by a sleep–wake behavior with a period length of 24.8 h and a diminished amplitude. Here, too, screening was performed by sequencing of candidate clock genes, resulting in the identification of a splice site mutation in the *CRY1* gene (rs184039278) that cosegregated with the trait. The mutation results in the deletion of exon 11 from the *CRY1* transcript, which in turn results in the deletion of 24 amino acids close to the C-terminal end of the protein. This turned out to be a gain-of-function mutation, which leads to enhanced interaction with the CLOCK and BMAL1 proteins, strengthening transcriptional inhibition in a way with is consistent with an increased period length.

Thus, we now have four examples of human clock gene mutations affecting the periodicity and, in one instance, the entrainment properties of the clock in a way that is in many ways more illustrative and dramatic than the knockout mouse models available for single clock genes. Taken together and combined with the evidence from hamster *tau* mutant, they form a beautiful illustration of the principle that circadian period length is regulated by the stability of the repressor complex (Takahashi, 2017). It is also evident that the direction of the effect—whether it slows down the oscillator by making the repressor more stable, or whether it speeds it up by making the repressor less stable—is dependent on the site and nature of the mutation rather than the protein involved. The hypothesis postulated in the infancy of mammalian clock gene research, based on previously established properties of the clock and the discovery of multiple repressor paralogs, of a separate oscillator based on *Per1* and *Cry1* that tracks dawn, and another one based on *Per2* and *Cry2* that tracks dusk, and that this would be reflected in the phenotypes of mutations in these genes (Daan et al., 2001), was elegant, but has turned out to be incorrect. What we are observing, instead, is how mutations in the components of the repressor complex can be loss-of-function or gain-of-function, determined by the functional consequences of the site of the mutation more than the general role of the mutated protein.

In spite of the huge increase in available whole genome and exome sequences over recent year, the first two reported mutations causing familial ASPD have not been reported in any other individuals worldwide. One interesting aspect of the most recently reported mutations, the *CRY2* mutation associated with ASPD and the *CRY1* mutation associated with DSPD, is the fact that they are, in fact, represented in the general population, at least among people of non-Finnish European ancestry. The ASPD allele of *CRY2* was not identified in the 1000 Genomes project, but has an allele frequency of 0.014% currently reported in the ExAc database. The DSPS allele of *CRY1* is represented only in groups of European ancestry components in the 1000 Genomes project, with allele frequencies of 0.5%–1.1%. In the

ExAc database, where it is represented in all categories except East Asians, the reported frequency is currently 0.4%. The highest allele frequency (0.6%) is found in individuals of non-Finnish European ancestry, and it cannot be excluded that the lower frequencies found in other population groups may be due to genetic admixture. The frequency in the general Turkish population was high enough for the authors of the study to be able to perform reverse phenotyping by investigating families with other homozygotic and heterozygotic carriers. In both homozygotes and heterozygotes, aberrant sleep behavior was observed, with late sleep timing and/or fragmented sleep (Patke et al., 2017).

7. GENETIC ASSOCIATIONS WITH MEASURED BEHAVIORAL RHYTHMS

The ideal scenario for mapping genetic associations with natural variation within the human clock is to have a precise quantitative measure rather than a quantitative or even qualitative self-reported trait. Such an ideal measure for community-based studies is activity as determined through actimetry, which is precise, economical, and noninvasive, albeit relatively time consuming. To date, there is only one such study to report, from the Harvard group which used 537 participants of European ancestry from the Rush Memory and Aging project (the MAP cohort) as their discovery cohort, and participants in their own inpatient circadian phenotyping studies as their control cohort (Lim et al., 2012). Using the acrophase of activity as their point of measure, they analyzed 135 polymorphisms within 18 candidate genes, and identified one (rs7221412) near the *PER1* gene, where homozygotes of the two alleles differed in timing by 67 min. A biological effect was also observed on the levels of *PER1* expression in cells collected from participants. This precise approach has much to recommend it, and is limited mostly by the fortnight or so it takes to collect actigraphy data from a participant, as opposed to spending a few minutes to answer a questionnaire. Consequently, the only genome-wide association studies (GWAS) of this type reported to date (Spada et al., 2016) failed to reach the consensus significance threshold, in all likelihood because of the low number of participants (956). Merging datasets from different locations may be problematic for the same reasons as MEQ scores from different locations are not directly comparable (von Schantz et al., 2015). Nonetheless, one would predict that the precision of this measure will yield interesting results in future.

8. SINGLE-GENE ASSOCIATION STUDIES OF QUESTIONNAIRE-DETERMINED HUMAN CHRONOTYPE

The normal or near-normal distribution of human chronotype, with heritability values ranging between 21% and 52%, as detailed in Section 3, is suggestive of a polygenic quantitative trait. A number of groups, including our own, have performed simultaneous collection of questionnaire data and DNA samples. While such collections are available with four-figure numbers of participants, they have not separately or jointly yielded any published significant GWAS data to date. Thus, data reported to date have been restricted to candidate gene analysis. A number of such associations have been reported, the most important ones of which are summarized later. An even larger number of associations reported between clock genes and other phenotypic traits fall outside of the scope of this review.

The first genetic association with chronotype to be reported was in the 5′-untranslated region of the *CLOCK* gene (rs1801260). It was suggested that carriers of the morningness allele (C) had an MEQ score on average 2.8 higher than T/T homozygotes (Katzenberg et al., 1998). However, with one exception (Mishima, Tozawa, Satoh, Saitoh, & Mishima, 2005), no other groups have reported being able to replicate this finding (Barclay et al., 2011; Chang, Buch, Bradstreet, Klements, & Duffy, 2011; Choub et al., 2011; Kim, Lee, Lee, Hwang, & Suh, 2016; Pedrazzoli et al., 2007; Robilliard et al., 2002). There is also a single report of association between an intronic polymorphism (rs922270) in the *ARNTL2* gene, which encloses BMAL2, and MEQ score (Parsons et al., 2014).

Studies on the genes encoding the repressor complex associations with extremes of MEQ score of a silent polymorphism in the *PER1* gene (rs2735611) (Carpen, von Schantz, Smits, Skene, & Archer, 2006) and a polymorphism in the 5'UTR of the *PER2* gene (rs2304672) (Carpen, Archer, Skene, Smits, & von Schantz, 2005), which had previously also been reported to associate with ASPD (Satoh, Mishima, Inoue, Ebisawa, & Shimizu, 2003). Two missense polymorphisms in *PER2* (rs934945) have been reported to associate with chronotype, one in two Korean population samples (Lee et al., 2011; Song et al., 2016) and another (rs2304671) in a Japanese population (Matsuo, Shino, Yamada, Ozeki, & Okawa, 2007).

The evidence about the most widely studied genetic association with chronotype, with the *PER3* gene, has (rather surprisingly) ended up suggesting that this gene has little if anything to do with the circadian clock dimension of the MEQ score. The VNTR polymorphism in the *PER3* gene (rs57875989) is located within the coding region, resulting in an 18-amino-acid motif being repeated (uniquely in primates Jenkins, Archer, & von Schantz, 2005) four or five times (Archer et al., 2003; Ebisawa et al., 2001). The longer, minor, and evolutionarily more recent 5-repeat allele was associated with greater morningness and appeared to be protective against DSPD (Archer et al., 2003). The association with morningness, as determined by the MEQ questionnaire (but not the suggestion of protection against DSPD), has subsequently been repeated in a number of different population samples (Kunorozva, Stephenson, Rae, & Roden, 2012; Lazar et al., 2012; Pereira et al., 2005) although not in some others (Barclay et al., 2011; Osland, Bjorvatn, Steen, & Pallesen, 2011). Nonetheless, a prospective study of homozygotes for the two different alleles rejected the hypothesis that the polymorphism was associated with different periodicity, as no differences were observed in any rhythmic parameters (Viola et al., 2007). Rather, the polymorphism has an effect on sleep homeostasis, such that homozygotes for the 5-repeat had a shorter sleep latency, a higher proportion of slow-wave sleep, and greater cognitive decrement during sleep deprivation. In other words, they behaved as if they were subject to a higher sleep pressure, and likely responded to questions in the manner of a morning type more because they were feeling more tired in the evenings. Thus, this polymorphism appeared to associate with the sleep homeostatic dimension of the MEQ, rather than the circadian one.

9. GWAS OF CIRCADIAN PHENOTYPE

As with a large number of other phenotypes, the development of GWAS, combined with the emergence of large cohorts of genotyped individuals with phenotypic information, is having a revolutionary impact in this field of research as well. GWAS of chronotype have emerged from two such large cohorts, the users of the home genetic testing product 23andMe (Hu et al., 2016) and the UK Biobank (Jones et al., 2016; Lane et al., 2016). Neither of these studies used a validated phenotyping instrument, relying instead on a single much less sophisticated basic self-characterization question, which in the case of 23andMe delivers a binary variable, while the UK Biobank questionnaire offered five answers including definite and moderate morning and evening types and intermediates. However, the imprecision of the phenotype has been amply compensated for by the sheer size of the cohorts, which are two orders of magnitude larger than those available to previous candidate gene studies. The results from these three studies have been fully integrated in a recent systematic review (Kalmbach et al., 2017) and will therefore only be summarized here. Significant hits within four genes, *PER2*, *FBXL13*, *RGS16*, and *AK5*, were found in all three analyses. The reappearance of *PER2*, a key component of the repressor complex in the circadian oscillator machinery, is a satisfying validation of the use of this key clock gene for candidate gene analysis—as detailed earlier, it was found to be mutated in familial ASPD (Toh et al., 2001), and a number of *PER2* polymorphisms have been implicated as diurnal preference alleles. *FBXL13* is a protein ubiquitin ligase (Curtiss et al., 2005). While it is not one of the F-box proteins previously implicated in the degradation of the repressor complex, this finding certainly identifies it as a strong candidate for further mechanistic study. *RGS16* had previously been shown to play a role in intracellular G-protein signaling within the SCN (Doi et al., 2011). The *AK5* gene, previously not associated with the circadian system, encodes a brain-specific adenylate kinase (Van Rompay, Johansson, & Karlsson, 1999). With a more general role in energetics and metabolic signaling, it is more difficult to extrapolate its possible specific role in the determination of chronotype.

Among GWAS hits not shared between all three studies, the only candidate that has previously occurred in this review is *PER3* (Hu et al., 2016). As previously mentioned, this gene appears to be involved in the sleep homeostatic rather than the circadian dimension of chronotype. The appearance of other genes with functions related to sleep, namely, *HCRTR2* (encoding a hypocretin receptor), *HTR69* (encoding a serotonin receptor) would seem to suggest that chronotype determined by a simple question probably captures a sleep homeostatic dimension as well as a circadian one.

These very promising GWAS have confirmed some genes already of interest for human circadian rhythms, and also identified a number of entirely novel ones. The specific SNPs pinpointed by the analysis are to be viewed as mapping positions of a resolution that is able to identify specific genes, but are very much more likely to be linked to the polymorphisms causing the phenotypic differences rather than being the causative ones themselves. Thus, these results identify novel research needs—both to characterize the phenotypic differences associated with these polymorphisms in greater detail, and to seek to identify the linked polymorphisms that cause these differences at the molecular level. In time, they are likely to be followed by further GWAS based on more refined phenotypes, such as chronotypes based on full questionnaires and precisely measured parameters collected through actigraphy.

10. SUGGESTIONS OF POPULATION DIFFERENCES IN HUMAN CLOCKS

The GWAS performed to date have all been performed exclusively in individuals of European ancestry. The reason for this was to avoid bias due to population structure. While many of the previous candidate gene studies referred to above did not explicitly restrict themselves to specific ethnic groups, in most cases there was a likely bias toward people of European ancestry because of the population structure in the countries in which they were performed. Differences between groups of different ancestries would be expected in general because of population stratification, and specifically because of the different photoperiod histories of different populations caused by large-scale historical population migrations as described earlier. The ancestors of modern Europeans have lived there for a few hundred generations. It is reasonable to hypothesize that these populations may have evolved circadian traits that allow them greater flexibility to adapt to high-amplitude photoperiods than African populations whose ancestors have never lived far from the equator. Natives of North and South America, by contrast, may be adapted to their current latitudes, but may also retain elements of their ancestral adaptation to the photoperiod of the area around Bering's Strait.

Differences in circadian parameters between human populations are fascinating, and an important potential contributor to health disparities between racial/ethnic groups (Egan, Knutson, Pereira, & von Schantz, 2017), but understudied. As is the case for practically any other human phenotype, the majority of published observations have been made on individuals of European descent. Added to this is the high cost of circadian laboratory studies, which many countries do not have resources for, the complicated cross-cultural features of chronotype distribution (von Schantz et al., 2015), and the fact that modern African-American and Native American populations are heavily admixed. Nonetheless, the data presented to date are intriguing. Work by Eastman and colleagues has suggested that African-Americans have shorter period length, larger phase advances, and smaller phase delays than European-Americans (Eastman, Molina, Dziepak, & Smith, 2012; Smith, Burgess, Fogg, & Eastman, 2009), even when controlling for genetic admixture (Eastman, Suh, Tomaka, & Crowley, 2015). Our own group utilized a behaviorally homogeneous population with a high degree of admixture to disentangle ancestral components to the MEQ score of chronotype. We observed no significant difference between European and ancestry, but a modest but significant association between Amerindian ancestry and morningness (Egan, Campos Santos, et al., 2017). These data suggest strongly that there are additional dimensions to the natural variation in human clocks and their genetic basis that we still know very little about. We have a substantial understanding about natural variation in human clocks, and have made important progress toward understanding its molecular determinants, based mostly on data collected from participants of European descent. We know that there are population differences in the distribution of clock gene alleles (Ciarleglio et al., 2008; Nadkarni, Weale, von Schantz, & Thomas, 2005). The recently reported alleles in *CRY1* for familial DSPD (Patke et al., 2017) and in *CRY2* for familial ASPD (Hirano et al., 2016) are absent in some population groups and may both in fact be restricted to individuals of European descent. In short, there is evidence for variability both in circadian phenotype and in the genes that create this variability. As we are beginning to see a clearer picture of how these relate to each other in Europeans and their descendants, an important part of the future research agenda must be to collect more data from people of different ancestries.

REFERENCES

Adan, A., Archer, S. N., Hidalgo, M. P., Di Milia, L., Natale, V., & Randler, C. (2012). Circadian typology: A comprehensive review. *Chronobiology International*, *29*(9), 1153–1175.

American Association of Sleep Medicine. (2014). *International classification of sleep disorders* (3rd ed.). Darien, IL: American Association of Sleep Medicine (ICSD-3).

Ancoli-Israel, S., Schnierow, B., Kelsoe, J., & Fink, R. (2001). A pedigree of one family with delayed sleep phase syndrome. *Chronobiology International*, *18*(5), 831–840.

Archer, S. N., Carpen, J. D., Gibson, M., Lim, G. H., Johnston, J. D., Skene, D. J., et al. (2010). Polymorphism in the PER3 promoter associates with diurnal preference and delayed sleep phase disorder. *Sleep*, *33*(5), 695–701.

Archer, S. N., Robilliard, D. L., Skene, D. J., Smits, M., Williams, A., Arendt, J., et al. (2003). A length polymorphism in the circadian clock gene Per3 is linked to delayed sleep phase syndrome and extreme diurnal preference. *Sleep*, *26*(4), 413–415.

Aschoff, J. (1958). Tierische Periodik under dem Einfluß von Zeitgebern. *Zeitschrift für Tierpsychologie*, *15*, 1–30.

Aschoff, J. (1960). Exogenous and endogenous components in circadian rhythms. *Cold Spring Harbor Symposia on Quantitative Biology*, *25*, 11–28.

Aschoff, J. (1979). Circadian rhythms: Influences of internal and external factors on the period measured in constant conditions. *Zeitschrift für Tierpsychologie*, *49*(3), 225–249.

Aschoff, J., & Wever, R. (1962). Spontanperiodik des Menschen bei Ausschluß aller Zeitgeber. *Naturwissenschaften*, *49*, 337–342.

Bae, K., Jin, X., Maywood, E. S., Hastings, M. H., Reppert, S. M., & Weaver, D. R. (2001). Differential functions of *mPer1*, *mPer2*, and *mPer3* in the SCN circadian clock. *Neuron*, *30*(2), 525–536.

Barclay, N. L., Eley, T. C., Buysse, D. J., Archer, S. N., & Gregory, A. M. (2010). Diurnal preference and sleep quality: Same genes? A study of young adult twins. *Chronobiology International*, *27*(2), 278–296. https://doi.org/10.3109/07420521003663801.

Barclay, N. L., Eley, T. C., Mill, J., Wong, C. C., Zavos, H. M., Archer, S. N., et al. (2011). Sleep quality and diurnal preference in a sample of young adults: Associations with 5HTTLPR, PER3, and CLOCK 3111. *American Journal of Medical Genetics Part B, Neuropsychiatric Genetics: The Official Publication of the International Society of Psychiatric Genetics*, *156B*(6), 681–690. https://doi.org/10.1002/ajmg.b.31210.

Brown, S. A., Ripperger, J., Kadener, S., Fleury-Olela, F., Vilbois, F., Rosbash, M., et al. (2005). PERIOD1-associated proteins modulate the negative limb of the mammalian circadian oscillator. *Science*, *308*(5722), 693–696. https://doi.org/10.1126/science.1107373.

Bunger, M. K., Wilsbacher, L. D., Moran, S. M., Clendenin, C., Radcliffe, L. A., Hogenesch, J. B., et al. (2000). Mop3 is an essential component of the master circadian pacemaker in mammals. *Cell*, *103*(7), 1009–1017.

Burgess, H. J., & Fogg, L. F. (2008). Individual differences in the amount and timing of salivary melatonin secretion. *PloS One*, *3*(8), e3055. https://doi.org/10.1371/journal.pone.0003055.

Campbell, S. S., & Murphy, P. J. (2007). Delayed sleep phase disorder in temporal isolation. *Sleep*, *30*(9), 1225–1228.

Carpen, J. D., Archer, S. N., Skene, D. J., Smits, M. G., & von Schantz, M. (2005). A single-nucleotide polymorphism in the 5′-untranslated region of the *hPER2* gene is associated with diurnal preference. *Journal of Sleep Research*, *14*, 293–297.

Carpen, J. D., von Schantz, M., Smits, M., Skene, D. J., & Archer, S. N. (2006). A silent polymorphism in the PER1 gene associates with extreme diurnal preference in humans. *Journal of Human Genetics*, *51*(12), 1122–1125.

Chang, A. M., Buch, A. M., Bradstreet, D. S., Klements, D. J., & Duffy, J. F. (2011). Human diurnal preference and circadian rhythmicity are not associated with the CLOCK 3111C/T gene polymorphism. *Journal of Biological Rhythms*, *26*(3), 276–279. https://doi.org/10.1177/0748730411402026.

Choub, A., Mancuso, M., Coppede, F., LoGerfo, A., Orsucci, D., Petrozzi, L., et al. (2011). Clock T3111C and Per2 C111G SNPs do not influence circadian rhythmicity in healthy Italian population. *Neurological Sciences*, *32*(1), 89–93. https://doi.org/10.1007/s10072-010-0415-1.

Ciarleglio, C. M., Ryckman, K. K., Servick, S. V., Hida, A., Robbins, S., Wells, N., et al. (2008). Genetic differences in human circadian clock genes among worldwide populations. *Journal of Biological Rhythms*, *23*(4), 330–340.

Crompton, A. W., Taylor, C. R., & Jagger, J. A. (1978). Evolution of homeothermy in mammals. *Nature*, *272*(5651), 333–336.

Curtiss, N. P., Bonifas, J. M., Lauchle, J. O., Balkman, J. D., Kratz, C. P., Emerling, B. M., et al. (2005). Isolation and analysis of candidate myeloid tumor suppressor genes from a commonly deleted segment of 7q22. *Genomics*, *85*(5), 600–607. https://doi.org/10.1016/j.ygeno.2005.01.013.

Czeisler, C. A., Duffy, J. F., Shanahan, T. L., Brown, E. N., Mitchell, J. F., Rimmer, D. W., et al. (1999). Stability, precision, and near-24-hour period of the human circadian pacemaker. *Science*, *284*(5423), 2177–2181.

Daan, S., Albrecht, U., van der Horst, G. T., Illnerova, H., Roenneberg, T., Wehr, T. A., et al. (2001). Assembling a clock for all seasons: Are there M and E oscillators in the genes? *Journal of Biological Rhythms*, *16*(2), 105–116.

Doi, M., Ishida, A., Miyake, A., Sato, M., Komatsu, R., Yamazaki, F., et al. (2011). Circadian regulation of intracellular G-protein signalling mediates intercellular synchrony and rhythmicity in the suprachiasmatic nucleus. *Nature Communications*, *2*, 327. https://doi.org/10.1038/ncomms1316.

Duffy, J. F., & Czeisler, C. A. (2002). Age-related change in the relationship between circadian period, circadian phase, and diurnal preference in humans. *Neuroscience Letters*, *318*(3), 117–120.

Duffy, J. F., Rimmer, D. W., & Czeisler, C. A. (2001). Association of intrinsic circadian period with morningness-eveningness, usual wake time, and circadian phase. *Behavioral Neuroscience*, *115*(4), 895–899.

Eastman, C. I., Molina, T. A., Dziepak, M. E., & Smith, M. R. (2012). Blacks (African Americans) have shorter free-running circadian periods than whites (Caucasian Americans). *Chronobiology International*, *29*(8), 1072–1077. https://doi.org/10.3109/07420528.2012.700670.

Eastman, C. I., Suh, C., Tomaka, V. A., & Crowley, S. J. (2015). Circadian rhythm phase shifts and endogenous free-running circadian period differ between African-Americans and European-Americans. Scientific reports, 5. Artn 8381, https://doi.org/10.1038/Srep08381.

Ebisawa, T., Uchiyama, M., Kajimura, N., Mishima, K., Kamei, Y., Katoh, M., et al. (2001). Association of structural polymorphisms in the human *period3* gene with delayed sleep phase syndrome. *EMBO Reports*, *2*(4), 342–346.

Egan, K. J., Campos Santos, H., Beijamini, F., Duarte, N. E., Horimoto, A. R. V. R., Taporoski, T. P., et al. (2017). Amerindian (but not African or European) ancestry is significantly associated with diurnal preference within an admixed Brazilian population. *Chronobiology International*, *34*(2), 269–272. https://doi.org/10.1080/07420528.2016.1265979.

Egan, K. J., Knutson, K. L., Pereira, A. C., & von Schantz, M. (2017). The role of race and ethnicity in sleep, circadian rhythms and cardiovascular health. *Sleep Medicine Reviews*, *2017*(33), 70–78.

Eide, E. J., Woolf, M. F., Kang, H., Woolf, P., Hurst, W., Camacho, F., et al. (2005). Control of mammalian circadian rhythm by CKIepsilon-regulated proteasome-mediated PER2 degradation. *Molecular and Cellular Biology*, *25*(7), 2795–2807. https://doi.org/10.1128/MCB.25.7.2795-2807.2005.

Elias, P. M., & Williams, M. L. (2015). Basis for the gain and subsequent dilution of epidermal pigmentation during human evolution: The barrier and metabolic conservation hypotheses revisited. *American Journal of Physical Anthropology*, *161*, 189–207.

Evans, D. S., Snitker, S., Wu, S. H., Mody, A., Njajou, O. T., Perlis, M. L., et al. (2011). Habitual sleep/wake patterns in the old order Amish: Heritability and association with non-genetic factors. *Sleep*, *34*(5), 661–669.

Gallego, M., Eide, E. J., Woolf, M. F., Virshup, D. M., & Forger, D. B. (2006). An opposite role for tau in circadian rhythms revealed by mathematical modeling. *Proceedings of the National Academy of Sciences of the United States of America*, *103*(28), 10618–10623. https://doi.org/10.1073/pnas.0604511103.

Gallego, M., & Virshup, D. M. (2007). Post-translational modifications regulate the ticking of the circadian clock. *Nature Reviews Molecular Cell Biology*, *8*(2), 139–148. https://doi.org/10.1038/nrm2106.

Gotter, A. L., Manganaro, T., Weaver, D. R., Kolakowski, L. F., Jr., Possidente, B., Sriram, S., et al. (2000). A time-less function for mouse timeless. *Nature Neuroscience*, *3*(8), 755–756.

Hasan, S., Santhi, N., Lazar, A. S., Slak, A., Lo, J., von Schantz, M., et al. (2012). Assessment of circadian rhythms in humans: Comparison of real-time fibroblast reporter imaging with plasma melatonin. *FASEB Journal: Official Publication of the Federation of American*

Societies for Experimental Biology, 26(6), 2414–2423. https://doi.org/10.1096/fj.11-201699.
Hirano, A., Shi, G., Jones, C. R., Lipzen, A., Pennacchio, L. A., Xu, Y., et al. (2016). A cryptochrome 2 mutation yields advanced sleep phase in humans. *eLife, 5*, e16695. https://doi.org/10.7554/eLife.16695.
Hogenesch, J. B., Gu, Y. Z., Moran, S. M., Shimomura, K., Radcliffe, L. A., Takahashi, J. S., et al. (2000). The basic helix-loop-helix-PAS protein MOP9 is a brain-specific heterodimeric partner of circadian and hypoxia factors. *The Journal of Neuroscience, 20*(13), RC83.
Horne, J. A., & Östberg, O. (1976). A self-assessment questionnaire to determine morningness-eveningness in human circadian rhythms. *International Journal of Chronobiology, 4*(2), 97–110.
Hu, Y., Shmygelska, A., Tran, D., Eriksson, N., Tung, J. Y., & Hinds, D. A. (2016). GWAS of 89,283 individuals identifies genetic variants associated with self-reporting of being a morning person. *Nature Communications, 7*, 10448. https://doi.org/10.1038/ncomms10448.
Hublin, J. J., Ben-Ncer, A., Bailey, S. E., Freidline, S. E., Neubauer, S., Skinner, M. M., et al. (2017). New fossils from Jebel Irhoud, Morocco and the pan-African origin of *Homo sapiens*. *Nature, 546*(7657), 289–292. https://doi.org/10.1038/nature22336.
Jenkins, A., Archer, S. N., & von Schantz, M. (2005). Expansion during primate radiation of a variable number tandem repeat coding region of the circadian clock gene *Period3*. *Journal of Biological Rhythms, 20*, 470–472.
Jenkins, D. L., Davis, L. G., Stafford, T. W., Campos, P. F., Hockett, B., Jones, G. T., et al. (2012). Clovis age western stemmed projectile points and human coprolites at the paisley caves. *Science, 337*, 223–228.
Jones, C. R., Campbell, S. S., Zone, S. E., Cooper, F., DeSano, A., Murphy, P. J., et al. (1999). Familial advanced sleep-phase syndrome: A short-period circadian rhythm variant in humans. *Nature Medicine, 5*(9), 1062–1065.
Jones, S. E., Tyrrell, J., Wood, A. R., Beaumont, R. N., Ruth, K. S., Tuke, M. A., et al. (2016). Genome-wide association analyses in 128,266 individuals identifies new Morningness and sleep duration loci. *PLoS Genetics, 12*(8), e1006125. https://doi.org/10.1371/journal.pgen.1006125.
Kalmbach, D. A., Schneider, L. D., Cheung, J., Bertrand, S. J., Kariharan, T., Pack, A. I., et al. (2017). Genetic basis of chronotype in humans: Insights from three landmark GWAS. *Sleep, 40*(2), zsw048. https://doi.org/10.1093/sleep/zsw048.
Katzenberg, D., Young, T., Finn, L., Lin, L., King, D. P., Takahashi, J. S., et al. (1998). A *CLOCK* polymorphism associated with human diurnal preference. *Sleep, 21*(6), 569–576.
Kim, S. J., Lee, J. H., Lee, S. Y., Hwang, J. W., & Suh, I. B. (2016). No association of CLOCK 3111T/C poymorphism with diurnal preference and sleep quality in Korean adults. *Sleep and Biological Rhythms, 14*, 135–140.
Kleitman, N. (1939). *Sleep and wakefulness*. Chicago: University of Chicago Press.
Kunorozva, L., Stephenson, K. J., Rae, D. E., & Roden, L. C. (2012). Chronotype and PERIOD3 variable number tandem repeat polymorphism in individual sports athletes. *Chronobiology International, 29*(8), 1004–1010. https://doi.org/10.3109/07420528.2012.719966.
Lane, J. M., Vlasac, I., Anderson, S. G., Kyle, S. D., Dixon, W. G., Bechtold, D. A., et al. (2016). Genome-wide association analysis identifies novel loci for chronotype in 100,420 individuals from the UK Biobank. *Nature Communications, 7*, 10889. https://doi.org/10.1038/ncomms10889.
Lazar, A. S., Slak, A., Lo, J. C., Santhi, N., von Schantz, M., Archer, S. N., et al. (2012). Sleep, diurnal preference, health, and psychological well-being: A prospective single-

allelic-variation study. *Chronobiology International*, *29*(2), 131–146. https://doi.org/10.3109/07420528.2011.641193.

Lee, H. J., Kim, L., Kang, S. G., Yoon, H. K., Choi, J. E., Park, Y. M., et al. (2011). PER2 variation is associated with diurnal preference in a Korean young population. *Behavior Genetics*, *41*(2), 273–277. https://doi.org/10.1007/s10519-010-9396-3.

Lim, A. S., Chang, A. M., Shulman, J. M., Raj, T., Chibnik, L. B., Cain, S. W., et al. (2012). A common polymorphism near PER1 and the timing of human behavioral rhythms. *Annals of Neurology*, *72*(3), 324–334. https://doi.org/10.1002/ana.23636.

Liu, A. C., Tran, H. G., Zhang, E. E., Priest, A. A., Welsh, D. K., & Kay, S. A. (2008). Redundant function of REV-ERBalpha and beta and non-essential role for Bmal1 cycling in transcriptional regulation of intracellular circadian rhythms. *PLoS Genetics*, *4*(2), e1000023. https://doi.org/10.1371/journal.pgen.1000023.

Looby, P., & Loudon, A. S. (2005). Gene duplication and complex circadian clocks in mammals. *Trends in Genetics*, *21*(1), 46–53.

Matsuo, M., Shino, Y., Yamada, N., Ozeki, Y., & Okawa, M. (2007). A novel SNP in *hPer2* associates with diurnal preference in a healthy population. *Sleep and Biological Rhythms*, *5*, 141–145.

Menaker, M., Moreira, L. F., & Tosini, G. (1997). Evolution of circadian organization in vertebrates. *Brazilian Journal of Medical and Biological Research = Revista Brasileira de Pesquisas Medicas e Biologicas*, *30*(3), 305–313.

Micic, G., Lovato, N., Gradisar, M., Ferguson, S. A., Burgess, H. J., & Lack, L. C. (2016). The etiology of delayed sleep phase disorder. *Sleep Medicine Reviews*, *27*, 29–38. https://doi.org/10.1016/j.smrv.2015.06.004.

Mishima, K., Tozawa, T., Satoh, K., Saitoh, H., & Mishima, Y. (2005). The 3111T/C polymorphism of hClock is associated with evening preference and delayed sleep timing in a Japanese population sample. *American Journal of Medical Genetics Part B, Neuropsychiatric Genetics*, *133*(1), 101–104.

Mongrain, V., Carrier, J., & Dumont, M. (2006). Circadian and homeostatic sleep regulation in morningness-eveningness. *Journal of Sleep Research*, *15*(2), 162–166. https://doi.org/10.1111/j.1365-2869.2006.00532.x.

Nadkarni, N. A., Weale, M. E., von Schantz, M., & Thomas, M. G. (2005). Evolution of a length polymorphism in the human *PER3* gene, a component of the circadian system. *Journal of Biological Rhythms*, *20*(6), 490–499.

Ohno, S. (1970). *Evolution by gene duplication*. Berlin: Springer.

O'Neill, J. S., & Reddy, A. B. (2011). Circadian clocks in human red blood cells. *Nature*, *469*(7331), 498–503. https://doi.org/10.1038/nature09702.

Osland, T. M., Bjorvatn, B. R., Steen, V. M., & Pallesen, S. (2011). Association study of a variable-number tandem repeat polymorphism in the clock gene PERIOD3 and chronotype in Norwegian university students. *Chronobiology International*, *28*(9), 764–770. https://doi.org/10.3109/07420528.2011.607375.

Papazyan, R., Zhang, Y., & Lazar, M. A. (2016). Genetic and epigenomic mechanisms of mammalian circadian transcription. *Nature Structural & Molecular Biology*, *23*(12), 1045–1052. https://doi.org/10.1038/nsmb.3324.

Park, O. (1940). Nocturnalism: The development of a problem. *Ecological Monographs*, *10*, 486–536.

Parsons, M. J., Lester, K. J., Barclay, N. L., Archer, S. N., Nolan, P. M., Eley, T. C., et al. (2014). Polymorphisms in the circadian expressed genes PER3 and ARNTL2 are associated with diurnal preference and GNbeta3 with sleep measures. *Journal of Sleep Research*, *23*(5), 595–604. https://doi.org/10.1111/jsr.12144.

Patke, A., Murphy, P. J., Onat, O. E., Krieger, A. C., Ozcelik, T., Campbell, S. S., et al. (2017). Mutation of the human circadian clock gene CRY1 in familial delayed sleep phase disorder. *Cell*, *169*(2) 203–215.e213. https://doi.org/10.1016/j.cell.2017.03.027.

Pedrazzoli, M., Louzada, F. M., Pereira, D. S., Benedito-Silva, A. A., Lopez, A. R., Martynhak, B. J., et al. (2007). Clock polymorphisms and circadian rhythms phenotypes in a sample of the Brazilian population. *Chronobiology International*, *24*(1), 1–8.

Pereira, D. S., Tufik, S., Louzada, F. M., Benedito-Silva, A. A., Lopez, A. R., Lemos, N. A., et al. (2005). Association of the length polymorphism in the human Per3 gene with the delayed sleep-phase syndrome: Does latitude have an influence upon it? *Sleep*, *28*(1), 29–32.

Posth, C., Wissing, C., Kitagawa, K., Pagani, L., van Holstein, L., Racimo, F., et al. (2017). Deeply divergent archaic mitochondrial genome provides lower time boundary for African gene flow into Neanderthals. *Nature Communications*, *8*, 16046. https://doi.org/10.1038/ncomms16046.

Preitner, N., Damiola, F., Lopez-Molina, L., Zakany, J., Duboule, D., Albrecht, U., et al. (2002). The orphan nuclear receptor REV-ERBalpha controls circadian transcription within the positive limb of the mammalian circadian oscillator. *Cell*, *110*(2), 251–260.

Preussner, M., & Heyd, F. (2016). Post-transcriptional control of the mammalian circadian clock: Implications for health and disease. *Pflugers Archiv - European Journal of Physiology*, *468*(6), 983–991. https://doi.org/10.1007/s00424-016-1820-y.

Ralph, M. R., Foster, R. G., Davis, F. C., & Menaker, M. (1990). Transplanted suprachiasmatic nucleus determines circadian period. *Science*, *247*(4945), 975–978.

Reick, M., Garcia, J. A., Dudley, C., & McKnight, S. L. (2001). NPAS2: An analog of clock operative in the mammalian forebrain. *Science*, *293*(5529), 506–509.

Richter, D., Grun, R., Joannes-Boyau, R., Steele, T. E., Amani, F., Rue, M., et al. (2017). The age of the hominin fossils from Jebel Irhoud, Morocco, and the origins of the middle stone age. *Nature*, *546*(7657), 293–296. https://doi.org/10.1038/nature22335.

Robilliard, D., Archer, S. N., Arendt, J., Lockley, S. W., Hack, L. M., English, J., et al. (2002). The 3111*Clock* gene polymorphism is not associated with sleep and circadian rhythmicity in phenotypically characterized human subjects. *Journal of Sleep Research*, *11*, 305–312.

Roenneberg, T., Kuehnle, T., Pramstaller, P. P., Ricken, J., Havel, M., Guth, A., et al. (2004). A marker for the end of adolescence. *Current Biology*, *14*(24), R1038–R1039.

Roenneberg, T., Wirz-Justice, A., & Merrow, M. (2003). Life between clocks: Daily temporal patterns of human chronotypes. *Journal of Biological Rhythms*, *18*(1), 80–90.

Rogers Ackermann, R., Mackay, A., & Arnold, M. L. (2015). The hybrid origin of "modern" humans. *Evolutionary Biology*, *43*, 1–11.

Satoh, K., Mishima, K., Inoue, Y., Ebisawa, T., & Shimizu, T. (2003). Two pedigrees of familial advanced sleep phase syndrome in Japan. *Sleep*, *26*(4), 416–417.

Shi, S., Hida, A., McGuinness, O. P., Wasserman, D. H., Yamazaki, S., & Johnson, C. H. (2010). Circadian clock gene Bmal1 is not essential; functional replacement with its paralog, Bmal2. *Current Biology*, *20*(4), 316–321. https://doi.org/10.1016/j.cub.2009.12.034.

Shirogane, T., Jin, J., Ang, X. L., & Harper, J. W. (2005). SCFbeta-TRCP controls clock-dependent transcription via casein kinase 1-dependent degradation of the mammalian period-1 (Per1) protein. *The Journal of Biological Chemistry*, *280*(29), 26863–26872. https://doi.org/10.1074/jbc.M502862200.

Smale, L., Lee, T., & Nunez, A. A. (2003). Mammalian diurnality: Some facts and gaps. *Journal of Biological Rhythms*, *18*(5), 356–366.

Smith, M. R., Burgess, H. J., Fogg, L. F., & Eastman, C. I. (2009). Racial differences in the human endogenous circadian period. *PLoS One*, *4*(6), e6014. https://doi.org/10.1371/journal.pone.0006014.

Song, H. M., Cho, C. H., Lee, H. J., Moon, J. H., Kang, S. G., Yoon, H. K., et al. (2016). Association of CLOCK, ARNTL, PER2, and GNB3 polymorphisms with diurnal preference in a Korean population. *Chronobiology International*, *33*(10), 1455–1463. https://doi.org/10.1080/07420528.2016.1231199.

Spada, J., Scholz, M., Kirsten, H., Hensch, T., Horn, K., Jawinski, P., et al. (2016). Genome-wide association analysis of actigraphic sleep phenotypes in the LIFE adult study. *Journal of Sleep Research*, *25*(6), 690–701. https://doi.org/10.1111/jsr.12421.
Taillard, J., Philip, P., Coste, O., Sagaspe, P., & Bioulac, B. (2003). The circadian and homeostatic modulation of sleep pressure during wakefulness differs between morning and evening chronotypes. *Journal of Sleep Research*, *12*(4), 275–282.
Takahashi, J. S. (2017). Transcriptional architecture of the mammalian circadian clock. *Nature Reviews Genetics*, *18*(3), 164–179. https://doi.org/10.1038/nrg.2016.150.
Tauber, E., Last, K. S., Olive, P. J., & Kyriacou, C. P. (2004). Clock gene evolution and functional divergence. *Journal of Biological Rhythms*, *19*(5), 445–458.
Toh, K. L., Jones, C. R., He, Y., Eide, E. J., Hinz, W. A., Virshup, D. M., et al. (2001). An *hPer2* phosphorylation site mutation in familial advanced sleep phase syndrome. *Science*, *291*(5506), 1040–1043.
Tucci, S., & Akey, J. M. (2016). Population genetics: A map of human wanderlust. *Nature*, *538*(7624), 179–180. https://doi.org/10.1038/nature19472.
van der Horst, G. T., Muijtjens, M., Kobayashi, K., Takano, R., Kanno, S., Takao, M., et al. (1999). Mammalian *Cry1* and *Cry2* are essential for maintenance of circadian rhythms. *Nature*, *398*(6728), 627–630.
Van Rompay, A. R., Johansson, M., & Karlsson, A. (1999). Identification of a novel human adenylate kinase. cDNA cloning, expression analysis, chromosome localization and characterization of the recombinant protein. *European Journal of Biochemistry*, *261*(2), 509–517.
Viola, A. U., Archer, S. N., James, L. M., Groeger, J. A., Lo, J. C., Skene, D. J., et al. (2007). *PER3* polymorphism predicts sleep structure and waking performance. *Current Biology*, *17*, 613–618.
von Schantz, M. (1998). The relationship between the retina and the circadian clock in mammals. *Medical Science Monitor*, *4*(Suppl. 1), 35–45.
von Schantz, M., Jenkins, A., & Archer, S. N. (2006). Evolutionary history of the vertebrate *Period* genes. *Journal of Molecular Evolution*, *62*, 701–707.
von Schantz, M., Taporoski, T. P., Horimoto, A. R., Duarte, N. E., Vallada, H., Krieger, J. E., et al. (2015). Distribution and heritability of diurnal preference (chronotype) in a rural Brazilian family-based cohort, the Baependi study. *Scientific Reports*, *5*, 9214. https://doi.org/10.1038/srep09214.
Welsh, D. K., Takahashi, J. S., & Kay, S. A. (2010). Suprachiasmatic nucleus: Cell autonomy and network properties. *Annual Review of Physiology*, *72*, 551–577. https://doi.org/10.1146/annurev-physiol-021909-135919.
Wittmann, M., Dinich, J., Merrow, M., & Roenneberg, T. (2006). Social jetlag: Misalignment of biological and social time. *Chronobiology International*, *23*(1–2), 497–509. https://doi.org/10.1080/07420520500545979.
Xu, Y., Padiath, Q. S., Shapiro, R. E., Jones, C. R., Wu, S. C., Saigoh, N., et al. (2005). Functional consequences of a CKIdelta mutation causing familial advanced sleep phase syndrome. *Nature*, *434*(7033), 640–644.
Xu, Y., Toh, K. L., Jones, C. R., Shin, J. Y., Fu, Y. H., & Ptacek, L. J. (2007). Modeling of a human circadian mutation yields insights into clock regulation by PER2. *Cell*, *128*(1), 59–70.
Zavada, A., Gordijn, M. C., Beersma, D. G., Daan, S., & Roenneberg, T. (2005). Comparison of the Munich chronotype questionnaire with the Horne-Ostberg's morningness-eveningness score. *Chronobiology International*, *22*(2), 267–278.
Zee, P. C., Attarian, H., & Videnovic, A. (2013). Circadian rhythm abnormalities. *Continuum*, *19*(1), 132–147.
Zhang, L., Hirano, A., Hsu, P. K., Jones, C. R., Sakai, N., Okuro, M., et al. (2016). A PERIOD3 variant causes a circadian phenotype and is associated with a seasonal mood trait. *Proceedings of the National Academy of Sciences of the United States of America*, *113*(11), E1536–1544. https://doi.org/10.1073/pnas.1600039113.

FSC
www.fsc.org
MIX
Papier | Fördert
gute Waldnutzung
FSC® C083411

Kolibri 360 GmbH
Ferdinand-Jühlke-Straße 7
99095 Erfurt, Deutschland
produktsicherheit@kolibri360.de